Pelican Books
ATOMS AND THE UNIVERSE

G. O. Jones was, until 1968, a professional physicist who specialized in solid-state and low-temperature physics. He has worked in the Universities of Oxford, Sheffield and London, and was for fifteen years Professor of Physics in the University of London. His publications include *Glass* and three novels. Since 1968 he has been Director of the National Museum of Wales.

J. Rotblat, C.B.E., is Professor of Physics in the University of London at St Bartholomew's Hospital Medical College. He is also the Secretary-General of the Pugwash Conferences on Science and World Affairs. He has been President of the British Institute of Radiology and the Hospital Physicists' Association, and the editor of *Physics in Medicine and Biology*. His publications include *Atomic Energy: A Survey* (1954), *Aspects of Medical Physics* (1966), and *Scientists in the Quest for Peace* (1972).

G. J. Whitrow is Professor of the History and Applications of Mathematics in the University of London at the Imperial College of Science and Technology. He has been President of a number of learned scientific societies and is currently President of the British Society for the History of Mathematics. His publications include *The Structure and Evolution of the Universe* (1959), *The Natural Philosophy of Time* (1961), and *What is Time?* (1972).

G. O. Jones
J. Rotblat
G. J. Whitrow

ATOMS AND
THE UNIVERSE

Third Revised Edition

Penguin Books

Penguin Books Ltd, Harmondsworth,
Middlesex, England
Penguin Books Inc., 7110 Ambassador Road,
Baltimore, Maryland 21207, U.S.A.
Penguin Books Australia Ltd, Ringwood,
Victoria, Australia

First published by Eyre & Spottiswoode 1956
Published in Penguin Books 1973

Made and printed in Great Britain by
Richard Clay (The Chaucer Press) Ltd,
Bungay, Suffolk
Set in Monotype Times

Contents

Plates

PLATES

Figures

Tables

Acknowledgements

Plate I (from *The Physical Review*) is reproduced by courtesy of the American Institute of Physics; IX by courtesy of Prof. P. M. S. Blackett; VI, VII, VIII (from *Radiations from Radioactive Substances* by Rutherford, Chadwick and Ellis) Cambridge University Press; XI Prof. P. B. Hirsch; II Prof. P. G. Kruger; XIII, XIV, XVI, XVII, XIX–XXVI and XXIX the Mount Wilson and Palomar Observatories; III Prof. C. F. Powell; XV, XVIII, XXVII and XXVIII the Royal Astronomical Society; V Prof. E. Segre; IV Prof. J. Steinberger; XII Prof. C. A. Taylor.

The authors wish to thank Magda Whitrow for compiling the Index.

Preface to the Pelican Edition

It was the object of this book, first published in 1956, to provide the interested layman with a readable outline of the essential content of modern physics. Many of the topics discussed have an influence far beyond their scientific importance and are frequently heard of in the world's press; we tried to give them more nearly their true scientific importance. In this third edition we have adopted the same approach but have brought the book up to date.

To give a survey of modern physics in three hundred pages we have had to abandon one of the main characteristics of scientific method – its insistence upon rigour in argument and explanation – and to simplify the discussion as far as possible without sacrificing essential truth. Fortunately, our readers will be familiar with words like 'force', 'momentum', 'energy', 'wavelength', which constantly appear in the following pages. Only when it is essential to be more precise, or to introduce unfamiliar terms, have we insisted on more rigorous definition.

Our general scheme is almost the reverse of historical, for we begin with the smallest objects, atoms and their constituent particles, which have been studied by scientists only during the present century. We then proceed to explain how atoms cohere to produce matter – in its many forms – and to discuss its behaviour. This is the world of familiar phenomena which was the main study of scientists during the nineteenth century. Finally, we move out of the laboratory to consider the largest objects – the planets and stars. This is where physical science began. So, starting with atoms we end by discussing the universe. In attending to details at all these levels there is a risk of failing to see the wood for the trees. We have tried to avoid this by including some account of the important unifying principles in science, and of the characteristic features of the scientific approach. We have also tried to illustrate the way in which the working scientist – rather than the philosopher – regards such matters.

The reception given to earlier editions of this book by university teachers reminded us that even specialist scientists often feel a need for a simplified entry into branches of science other than their own. Indeed, our own discussions while writing the book showed this to be true, because each of us found himself to be something of a layman in the specialist fields of the others. We believe that this is a book for the interested layman, provided he is reasonably determined as well as interested. At the same time, because the range covered is wide, we believe that many students of science at school and university, and many working scientists, may also – like ourselves – find benefit from this book.

G. O. J.
J. R.
G. J. W.

Introduction

The Study of the Physical World

By the simple act of looking around him, inquiring man confronts the main problems faced by the scientist. For he is not satisfied merely to look; he very soon wants to know what the objects he sees are made of and how they behave. That is, he wishes to understand the structure, constitution and properties of matter. Next, he may ask how he sees objects. He is now asking what are the properties of light, how it is emitted or reflected from matter, and how it affects the retina of his eye. So far he has been asking questions which a physicist would try to answer. He will probably continue by asking how he, a particular kind of lump of matter, has the ability to translate the images on the retina into mental images, and indeed, to ask such questions. Ultimately he may ask how (or perhaps *why*) it is that he exists at all. He has already ventured into biology, metaphysics and philosophy.

In this book, while we aim to survey the physical world as a whole, we have discussed the facts and problems of physics at some length. Do we imply that physics covers the whole physical world? This is an important question, which it is worth pausing to examine, because this will give us at once a view of the content of science and of the inter-relations between its branches.

We admit at once that physics does not cover all the world in any conventional sense. Nor, indeed, would we claim that all the other well-known branches of science are in any way subordinate to physics. However, the borders between physics and astronomy, chemistry, biology or geology – to name a few of the most important sciences – are less real than most people imagine, and physics has a rather special and fundamental role to play in relation to these other sciences. Let us consider first physics, astronomy and chemistry. It will be necessary to say something about forces, and about the motions to which they give rise, in order to have a basis for discussion.

Force is one of the commonest themes in physics. Bodies can act upon each other with various kinds of force which are distinguishable from each other. For example, the gravitational force of attraction exists between all massive bodies. This force keeps the moon and planets in their orbits and causes bodies on earth to fall, although the gravitational force between small bodies can be detected only by very careful experiments. Electric forces between electric charges can be easily demonstrated by rubbing a piece of plastic on one's sleeve. This action causes loose electricity to appear on the plastic, which will then attract light objects such as small pieces of paper. Magnetic forces of attraction or repulsion such as exist between magnets, or between magnets and pieces of iron, are easily recognized.

Physics deals in detail with the nature of forces and with their effects. Take the case of a body falling to the earth under the action of the downward gravitational force of attraction between it and the earth; in the absence of air the speed of fall would become greater and greater indefinitely. Actually it becomes constant when the upward force of drag due to the air becomes as great as the downward gravitational force – when there is no net force acting on the body.

Physical 'laws' of this kind are constantly demonstrated and verified in elementary physics laboratories and exactly the same laws govern the motions of the stars and planets. Thus physics and astronomy are part of a continuum of knowledge: while astronomy deals with the motions of heavenly bodies, these motions are governed by laws which form one part of the province of physics. The part of this continuum which is essentially physics concerns the fundamental nature of the laws relating to forces and motions; the part which is essentially astronomy concerns the detailed motions of the stars and planets. Again, the study of the nature of light is one of the main branches of physics; the study of the light emitted by stars is one of the most powerful methods by which their natures can be found – and is, therefore, of great importance in astronomy. Several other examples illustrating the continuity between physics and astronomy will be mentioned in this book.

Let us consider next the relation between physics and chemistry. Chemistry is, essentially, the study of the differences between substances. Everyone knows that the substance water is represented by 'H_2O' even though he may not understand the significance of this symbol. It means that the smallest 'unit' of water which can exist – a molecule of water – consists of two atoms of hydrogen (represented by 'H_2') and one atom of oxygen (represented by 'O') stuck together. These hydrogen and oxygen atoms can be separated again by passing an electric current – the electrolysis of water. Conversely, the oxygen and hydrogen can be re-formed into water by an electric spark. A large body of knowledge has been built up over many centuries concerning the ways in which different substances can be formed and transformed. Hydrogen and oxygen are two of about a hundred different kinds of atoms which can behave as fundamental units and are accordingly known as elements. When atoms of two or more elements stick together (or 'combine') they form molecules of other substances known as compounds, of which water is an example.

With a hundred kinds of brick one might expect to be able to build very many different kinds of structure and in fact the number of chemical compounds which can exist is without limit: the lightest element, hydrogen, can combine with nearly every other element, and indeed, nearly every element can combine with nearly every other. Chemistry is thus a limitless subject in its own right. It is indeed extraordinary how far chemistry developed before the properties of atoms had been discovered. Their existence was tacitly assumed by chemists as necessary for the development of a logical pattern in their subject.

Just as physics deals with the laws governing the motions of the stars and planets, it also deals with fundamental aspects of the discoveries of chemistry, attempting to explain why some elements combine under certain circumstances or can be separated under others. It attempts to discover what is the 'glue' which sticks atoms together to form molecules. In this respect it has in the present century been brilliantly successful in spite of the enormous complexity of the problem. The discoveries made in physics about the nature of atoms have also shown, at least in principle, why atoms should stick together. One type of force is electrical. In water, for example, the atoms stick together largely because they become electrically charged. By the application of an electric force, in electrolysis, they can be drawn apart. At this point, where we have a physical process leading to chemical results, physics and chemistry are inseparable and indistinguishable.

Physicists often say to chemists that chemistry is a branch of physics – meaning that physicists study those aspects of chemistry which appear to be the more fundamental and leave to chemists the work on the details of particular cases. The chemist might reply that physics is a tool which he uses to solve his own problems. The impartial reader will perhaps be content to draw the conclusion that, at least, the two disciplines must be part of a continuous whole.

So far, we have mentioned phenomena which differ in scale rather than in kind. But where do the biological sciences stand? Is there not something different about sciences which deal with the study of life? The answer, in so far as these questions can be answered, may be somewhat surprising to the reader, for it contains a notable demonstration of the rather dispassionate attitude which natural science adopts even to such questions as 'What is life?' which are so significant in philosophy and metaphysics.

Biology has developed gradually from mere classification, through the emergence of general principles such as the theory of evolution, to a synthesis with physics and chemistry which is quite as important for its future as was the synthesis with physics for the development of chemistry. When the mechanisms of biological processes– such as cell division – are studied in ultimate detail they are found to be purely physical and chemical processes. Of course, the living processes are not those normally studied by physicists and chemists, but they are not fundamentally different and it would be quite possible to list a sequence of different processes, running from the ordinary chemical to the ordinary biological, in which each step was too small to represent a real difference. In a very large number of such steps, the difference between a man and a lump of metal could be encompassed and we do not, of course, assert that a man is the same kind of thing as a lump of metal. But in trying to answer the question 'What is life?' we are somewhat at a loss since it is hardly a sufficient answer to point to the two ends of our sequence and to say that life, or the absence of life, is what distinguishes them from each other. In this dilemma, the model dilemma of philosophy, it is useful to remember Humpty-Dumpty's assertion in *Alice through the Looking-glass*: 'When I use a word, it means what I choose it to mean, neither more nor less.' What we choose to regard as the boundary between living and non-living is purely a matter of convention.

Some of the latest developments in biology have resulted from the application of the newest ideas and techniques of physics and chemistry. Physics, for instance, has links with biology which depend on the application of physical techniques. The invention of the microscope was made possible by the discovery of the laws governing the reflection of light in passing through materials like glass – matters which were the typical concern of early physicists. More recently, we have seen the introduction of instruments – such as the electron microscope – which depend on some of the later discoveries of physics, as well as on many new methods of analysis developed by chemists, all used to provide new and powerful tools for the biologist.

We have seen that astronomy, physics, chemistry, biology are all part of a continuum of science. Mathematics too is a part of the continuum but also, as we illustrate in the Appendix, the servant of them all. Science is now so vast that the fraction of the continuum which can be occupied by one man is only tiny. No scientist is ashamed to declare his field quite narrowly by identifying himself as, say, a metal physicist, or a glass chemist, or to declare that his field lies at the border-line *between* two of the ancient sciences – say biomathematics or astrophysics.

The Methods of Experimental Science

All the content of science depends, in the last resort, on experimental science. Galileo's experiment, in which he rolled balls of different weights down an inclined plane, exemplifies one of the characteristic methods of approach in that it had a definite object, and yielded a definite result. On other occasions scientific discoveries seem to be almost accidental, although even then the scientist is usually looking for something. It is usually no accident that he is looking where he is, even if he does not know for what he is looking. The discovery of penicillin by Fleming is a well-known example of an apparently accidental success: a spore of a common green mould happened to settle on a culture plate upon which bacteria were under examination. Of course, it required a man of Fleming's acumen to realize the significance of the event. Equally, it required the most refined chemical techniques, and the skill and perseverance of Florey, Chain and their colleagues before the active substance could be extracted in the laboratory, many years later.

One of the most famous experiments of all time was the identification of the electron by J. J. Thomson in 1897. This experiment is most appropriate for more detailed discussion because it introduces many concepts which will later be useful to the reader. Thomson's experiment marks a turning-point in physics, since the electron, one of the fundamental particles of which atoms are made up, was the first such particle to be discovered. Today, electrons are used in all the modern devices of electronics, such as radio and television receivers, electronic computers, and devices for the automatic control of industrial processes and the automatic guiding of missiles.

It had been discovered earlier that, when an electric current was passed through a gas (in a discharge tube) under certain conditions, a beam of some kind travelled in the tube from the negative to the positive terminal, that is, in the direction opposite to that which had conventionally been accepted as the direction of flow of current. This beam of cathode rays could be made to cause scintillations on a suitable screen; an object placed between the terminal and the screen would then cast a shadow on the screen, indicating that the beam normally travelled in a straight line. A barrier pierced with a small hole could be used to select a narrow 'pencil' out of the main beam in order to show its path more clearly. It was already known as a result of the discoveries of the eighteenth and nineteenth centuries that electric charges could be caused to move by the application of electric forces and also that a wire carrying a current could be caused to move by the application of a magnetic force. In this case the direction of motion is at right angles both to the direction of the current and to that of the magnetic force. The physical laws governing these motions had indeed become an accepted part of physics. Now when electric or magnetic forces were applied at right angles across the path of the beam it was found that the beam was deflected *as a whole*. These results were sufficient in themselves to show that the beam consisted of a stream of negatively charged identical particles whose flow constituted at least part of the current.

J. J. Thomson made use of these deflections in order to establish for the individual particles the value of the quantity e/m – the ratio of the electric charge e to the mass m. This method consisted of applying a transverse electric force in order to deflect the beam and then finding what magnetic force applied in the other transverse direction would just return the beam to its original path. Knowing the magnitudes of the electric and magnetic forces applied he could obtain the value of e/m. We need not worry here about the mathematical analysis required in devising the experiment and in working out the result (see Appendix). But it is worth reminding ourselves that all that was *actually observed* was the point of impact of the beam, and that the experiment had to be designed to make this a useful observation.

The significance of the result was enormous. By showing that a definite value for e/m existed Thomson had really discovered the electron – as this new particle was called – even though its manifestations had already been seen in the observations which had led up to his experiment. It is usual to speak of this experiment as the identification of the electron, and this is perhaps just. Also, since the same value was obtained whatever gas was contained in the tube, the particle identified was clearly a sub-atomic particle –

that is, a constituent particle of atoms. From that time on, any other sub-atomic or new particle could be subjected to some similar test in order to see whether it was of the same kind.

The necessity for exact measurement of the kind used in this experiment is very characteristic of physics, and it is of interest to consider what would be involved in making the experiment accurate. The final answer depends on knowing the magnitudes of the electric forces and magnetic forces applied. In practice we always set up some convenient unit or standard before any physical quantity is measured – that is, before its magnitude can be given a number. The volt per centimetre is a convenient unit of electric force. If a 1-volt battery were connected across parallel plates 1 metre apart, the electric force between the plates would be 1 volt per metre. So in the Thomson experiment it is necessary first to make sure that the parallel plates really are quite parallel, and to measure the distance between them very accurately. Then the voltage across the plates is determined with a voltmeter. The reader will be familiar with instruments which do not necessarily read correctly – perhaps he has suffered from errors in the indication given by a petrol gauge on a car. In physics the question of accuracy in instruments is vital. The voltmeter has to be very well constructed in order to make sure that a given voltage always causes the same deflection of the pointer. At the time of manufacture it is compared with a standard instrument of known accuracy. And, most important, the standard instrument itself is made in such a way that it reads '1 volt' when subjected to the voltage defined as 1 volt. The definition of the unit known as the volt in itself implies that its measurement can be realized in practice. We have considered some of the experimental problems which are involved in measuring the electric force in the Thomson experiment. These are rather more easily explained than those associated with the magnetic force, but of course both have to be measured, and in a complex modern experiment very many more quantities might have to be measured accurately at the same time.

The next steps in the history of the electron were the attempts at measuring its mass, or its charge, separately. The first accurate experiment of this kind was performed by R. A. Millikan in 1909.

He found that minute oil droplets floating in a chamber and observed by a microscope could be made to pick up charges and could then be held suspended, or accelerated upward or downward, by an electric force applied between parallel plates. The charge, determined by measuring the rate of motion under a given electric force, always varied by multiples of a certain unit of charge. He presumed, as we now know correctly, that this 'highest common factor' was the charge of a single electron, and e was thus determined. By making use of this result in conjunction with Thomson's value of e/m it was then a simple matter to work out the value of m, the electronic mass. This turned out to be very much smaller than the masses of atoms; it is about one eighteen-hundredth part of the mass of the hydrogen atom.

In Millikan's experiment, perhaps more than in Thomson's, experimental skill and refinement of the highest order were necessary. For Thomson measured only the *ratio* between two quantities which were minutely small by comparison with the electric charges and masses normally studied in the laboratory. But Millikan actually observed the effects of single electrons, each equivalent to an electric charge of only 1.6×10^{-19} coulombs. The significance of this can be seen when one realized that the choice of the unit of charge is made arbitrarily in order to have a size convenient for ordinary experiments in the laboratory. (A coulomb is a charge which passes through a conductor when a current of 1 ampere flows for 1 second.) So Millikan observed the effects of charges which were smaller by a factor of about 10^{10} (or 10 000 000 000) than those which would normally have been examined in laboratory experiments on electricity – a feat of refined experimenting which has since been emulated in many of the experiments of physics.

Physics has travelled a long way since the experiments of Thomson and Millikan. We have already seen how a beam of electrons can be made to travel in a straight line and can also be deflected, or bent, just as a beam of light is bent in a lens or prism. The special significance of using electrons for microscopy is that they are very small, and can be made to 'see' things which are too small to be seen by the use of light (see Plate XI). For, just as the long radio waves can go around corners, the much shorter light waves fail to identify objects much shorter than their wavelength. The wavelength of ordinary visible light is still quite large by atomic standards – perhaps 1 000–5 000 times the diameter of an atom, so that objects consisting of hundreds or millions of atoms, which in linear dimension are only tens or hundreds of atomic sizes, are still too small to be seen in any optical microscope. In biology, in particular, many of the most important objects are of this kind of size.

In the pages which follow, the reader will be presented with experimental, observational and theoretical results derived from a multitude of researches. In general, there will be no opportunity to explain how these results were obtained, but we hope that these remarks will have helped to make the reader aware of the typical background behind the main conclusions of scientific research. Some further discussion of this background will be found in the Appendix.

The New Scientific Revolution

Although in this book we concentrate upon the pure content of science, and only occasionally refer to its methods or its applications, we cannot ignore the fact that the whole character and scale of scientific activity have changed since the Second World War. Most notable has been the dramatic increase in the *amount* of scientific activity (particularly in the application of science to practical problems), in the power and sophistication of the methods of science, and in public awareness of the implications of scientific results.

In the earlier history of science, research was carried out mostly by wealthy amateurs, or by university professors or college fellows. Their inspiration was the wish of the scholar to understand and find out as much as possible about the world. It became a part of the tradition of university life, especially in science faculties, that the teachers pursued pure research in their subjects in order to keep their own knowledge and interest lively and to stimulate, by example, those whom they taught. There was almost no interest in the possible applications of their results.

It was probably the undoubted success of physicists in the Second World War which caused the new scientific revolution.

Industrial organizations and governments, previously somewhat half-hearted in their interest, entered the field of scientific research decisively, and are now by far the largest employers of scientists. Their interest is of course mainly, though not entirely, in the applications of discoveries already made in pure science. But one of the lessons of war-time science was how quickly pure science can be turned into applied science. It was only six years after the discovery of nuclear fission that the first atomic bomb was dropped.

So the interest of industry and government in pure science is not entirely altruistic. And, of course, there have been innumerable demonstrations of the fact that research in applied science with short-sighted vision can be less successful in obtaining the desired result than apparently detached research in pure science. Faraday's pure researches at the Royal Institution led to methods for the generation of electrical power. It has often been said that if a government commission of the day had wished to improve methods of lighting, it would have awarded grants to those who wanted to work on the improvement of candles.

The discovery of nuclear fission was made by individuals, working with simple apparatus. The development of the atomic bomb was the work of thousands. But large-scale science is not always applied science. In some branches of science there has recently been the appearance of large 'team' projects which are nevertheless researches in pure science. The reason for putting a team of scientists to work on a problem is always the same: complexity. In many of the complex problems of modern biology, for example, it is difficult to make progress without calling in the services of physicists and chemists, and there now exist in many institutions teams consisting of scientists who might identify themselves separately under any of the following labels: biologist, biochemist, biophysicist, chemist, mathematician, physicist. At the same time they might be assisted by technicians who were expert, separately, in photography, electronics, computing, X-ray and many other techniques. In nuclear research pieces of apparatus have been constructed which cost millions of pounds, and are designed and run by large teams containing engineers, mathematicians and physicists. Such machines are under the control of pure scientists (although Government financial support is necessary), and are not remotely connected with the production of atomic bombs or fuel. The experiments carried out are, in principle, hardly different from the small-scale experiments carried out by J. J. Thomson in his laboratory, or by thousands of individual research workers all over the world. They are, indeed, logical developments of Thomson's experiment, and in the same tradition.

The border-line between pure and applied science is sometimes obscure. But they are now linked irrevocably in another way because of the powerful impact of the *techniques* of applied science, themselves derived from pure science, upon its own experimental methods. A number of examples of this will be mentioned in Chapter 7.

Looking broadly at the whole scene, it is astonishing how rapid has been the growth of scientific effort during one generation. It may be illustrated by saying that, whereas less than 5 per cent of all the human beings who have ever lived are now alive, 80 to 90 per cent of all the scientists who have ever lived are alive today. And although it might have been possible to make the same startling comparison at earlier times, it is only recently that the scientific explosion, so much more rapid than the population explosion, has overtaken it in the public consciousness. For the scientist himself science seems to be a young activity; most of it will have taken place during his lifetime, most of his colleagues will have been young men.

But there has been another – and more sombre – change in the role of science in public awareness, a loss of optimism, signified perhaps by the drift from science among school-children, and by the deliberate adoption of irrational creeds by many students. Not many young people now hold the Wellsian vision of science of, say, fifty years ago. It has been too grossly muddied by the events of those years. Who can deny that among the most obvious results of science and technology have been the nuclear arms race, environmental pollution and trash in the media of public entertainment?

It is not enough for the scientist to say that science is neutral. Indeed, we think that scientists have a responsibility for the social implications of their work. In this book, however, we present only the objective content of science, in its own beauty and internal harmony, revealed through endeavour, skill, genius and integrity. We believe that the story of science carries a message for humanity.

CHAPTER 1

Elementary Particles

The Atomists and the Alchemists

What is the structure of matter? Is there anything in common in the make-up of different substances, in a rose and a ruby; in a piece of brass and the human brain? If there is, what is the nature of the prime materials, of the simple bricks of which all matter is built? How many are there? Are they really the simplest forms of matter, or can they be still further divided? This sort of question engaged the mind and aroused the curiosity of man ever since he learned to think. Throughout the history of civilization one can discern a strong desire to solve the riddle of the structure of matter, to shed light on the nature of the elementary particles of which it is supposed to be composed. It was not until the twentieth century that real progress was made towards solving these problems but even now we have no complete answer to the above questions.

The early concepts of the structure of matter appear to have their origin in ancient Greek philosophy. In the fifth century B.C. the atomistic hypothesis was put forward by Leucippus and Democritus. According to them, all substances are built up of small units called atoms which are the smallest fragments into which a given substance can be divided. This is implied in the name 'atom' which means 'indivisible'. At about the same time a different hypothesis was put forward by Empedocles and later developed by Aristotle. They believed that all matter consists of one primordial substance called *hyle*, which is identical in all bodies. The difference between various substances is due to the presence in them in varying quantities of certain qualities imposed on the prime matter by the four elements, fire, earth, air and water.

The great authority of Aristotle was probably responsible for the general acceptance of this concept for over two thousand years. It formed the basis of alchemy, which in the Middle Ages was the forerunner of modern nuclear physics. The alchemists' aim was to transform elements, particularly to make noble metals from base ones. In order to achieve this they tried to obtain pure prime matter by taking away from a substance its four elements. They thought that once having achieved this the transmutation of elements could be easily accomplished by adding the four elements in the suitable proportions.

The complete failure of the alchemists brought to the foreground the long neglected atomistic concept, which became an established theory by the beginning of the nineteenth century, mainly due to the work of John Dalton. From consideration of the ways in which various elements combine together to form chemical compounds, Dalton arrived at the conclusion that every element is built up of atoms which are the indestructible and indivisible units of matter. The atoms of each chemical element are identical but different from atoms of other elements. Since we now know over a hundred different chemical elements, including those produced artificially (Fig. 1), it would follow from Dalton's theory that there are that number of elementary particles. This is not a very satisfactory conclusion. It is somewhat difficult to accept the idea that so many different types of brick have to be used in the structure of the universe. How much simpler and more attractive seemed the hypothesis, put forward by W. Prout in 1816, that the atoms of all elements are built up of one atom, the atom of hydrogen, which would thus correspond to the prime matter, the *hyle* of the Greeks. Prout based his idea on the assumption that the atomic weights of all elements were whole numbers and, therefore, multiples of the atomic weight of hydrogen, which is given unit value. This assumption proved to be incorrect when, stimulated by Prout's hypothesis, more accurate measurements of atomic weights were carried out and found in many cases to have fractional values, as for example 35·457 for chlorine, or 63·54 for copper. Prout's hypothesis was thus abandoned, but only temporarily, for it was revived in a modified form a century later.

The fact chiefly responsible for this revival was the discovery of radioactivity and the evidence it has brought that the atom is destructible. But even before this discovery a number of phenomena became known which had a decisive influence on our ideas of the structure of matter. These phenomena, which deal with the discharge of electricity in gases at low pressures, could only be observed following the development of vacuum techniques, i.e. of methods of producing low pressures, in the latter half of the nineteenth century.

The Electron

The passage of electricity through gases at low pressures was the subject of study by a number of scientists, notably by E. Goldstein in Germany and W. Crookes in Great Britain. If an electric potential difference of several thousand volts is applied to two terminals inserted at the ends of a glass tube, from which the air can be pumped out, it is found that at a sufficiently low pressure a green glow, or luminescence, appears in the tube. The cause of this luminescence can be traced to some radiation originating at the negative terminal, the cathode; for this reason the radiation was given the name 'cathode rays'.

The nature and properties of the cathode rays were fully investigated by J. J. Thomson, as already described in the Introduction. In these experiments the existence of a particle much lighter than the lightest of all atoms was established for the first time. This particle was given the name 'electron', and we know now that its mass is 1,837 times smaller than that of the hydrogen atom, and that the negative electric charge it carries is the smallest that can be observed; in fact, every electric charge so far observed is an integral multiple of the charge of the electron.

An important fact revealed in these studies was that, although the velocity of the cathode rays was variable, depending on the voltage applied to the tube, the charge and the mass of the electrons was always the same, if a correction was made for the variation of mass with velocity which follows from the relativity theory (see Chapter 6). They remained constant and independent of the nature of the gas in the tube or of the material of the electrodes.

3

Cathode rays are not the only means of producing electrons. In fact, three other phenomena in which electrons are emitted were discovered at about the same time. One was the so-called 'photo-electric effect', which has nowadays many practical uses, notably in television. It occurs when certain substances, particularly the alkali metals, are irradiated with visible light, or still better with ultra-violet light. A stream of particles is then found to issue from the metal, and these particles were identified as electrons. Another effect is the so-called 'thermionic emission', which is the basis of the functioning of radio valves. If a metal filament is heated to a high temperature, charged particles are emitted from it which can produce fluorescence on a screen, as in a television tube, or cause an electric current to flow in a circuit. These particles too were identified as electrons. Finally, it was found that electrons were emitted spontaneously from certain elements, the radioactive substances. These are very fast electrons, and before they were identified as such were given the name of β-rays.

The amazing fact is that although produced in such a variety of ways, all these rays have the same properties, the same electric charge and the same mass. No matter whether they are produced in solids or in gases, at low or high temperatures, at atmospheric pressure or in vacuum, by irradiation with light or by spontaneous emission, they are all identical. The obvious conclusion is that the electrons go into the make-up of all elements and that, therefore, they are a universal constituent of matter. It has proved impossible so far to produce a charge smaller than that carried by an electron, or to break it up into smaller fragments. The electron may thus be considered to be an elementary particle.

The Proton

Although electrons have proved to be ingredients of all elements they cannot be the only constituents of matter. For one thing, electrons are negatively charged, while atoms are normally neutral; there must, therefore, exist particles of positive charge. Moreover, the mass of the electron is far too small to account for the mass of the atom. The clue to the identity of the other elementary particles came as a result of the discovery of radioactivity.

The subject of radioactivity will be discussed in detail in the next chapter, but a general outline will be useful here. Thanks to the work of Becquerel in 1896, and of Marie and Pierre Curie in the following years, it was established that the heaviest elements existing in nature, such as uranium, thorium, radium or polonium, have the property of sending out radiations spontaneously. These radiations are known as a-, β- and γ-rays. As has already been stated, β-rays are fast-moving electrons; a-rays were found to be positively charged heavier particles, and γ-rays were shown to be electromagnetic waves. The most revolutionary discovery made in radioactivity was that after the emission of a- or β-rays the atoms of a given radioactive substance are transformed into atoms of a different chemical element; in many cases these are also radioactive and in turn undergo a further transformation. By 1902 Rutherford and Soddy had established the laws governing the transformation of radioactive elements and had shown how the various radioactive substances can be arranged in three families (Fig. 6), which start with uranium and thorium and end with lead after a whole series of transformations. Here, for the first time, the dream of the alchemist, the transmutation of elements, came true. It was found to be a spontaneous process which had been going on undetected for ages. Obviously, if an atom can break up and change into another, then it cannot be considered an elementary particle. The discovery of radioactivity thus brought to an end the belief that atoms are indivisible and elementary particles of matter.

Further experiments with radioactive substances, mainly the investigations on the scattering of a-particles made by Rutherford, gave an intimation of the structure of the atom. From these experiments, and the theory put forward by Niels Bohr in 1913, the nuclear model of the atom was evolved. According to this, each atom resembles a solar system; it is built up of a central core, or nucleus, and of a number of electrons revolving around it at various distances. The nucleus is positively charged, and the number of electrons is such that their total charge exactly balances the positive charge of the nucleus so that the atom as a whole is electrically neutral. The number of electrons revolving round the nucleus in an atom has proved to be a very important property of the atom. It defines the numerical place occupied by the given element in the Periodic Table of elements (Fig. 1); for this reason it is called the 'atomic number'. Thus, hydrogen, which has one electron in its atom, occupies the first place in the Table; helium with two electrons, the second place; oxygen the eighth; uranium the ninety-second. The atomic number also determines the chemical properties of the given element, since all chemical processes are caused by changes in the configuration of the electrons; the nucleus itself does not take any direct part in these processes.

Rutherford's scattering experiments also revealed that the nucleus is extremely small, its radius is about ten thousand times smaller than that of the atom, but despite its minuteness it contains practically all the mass of the atom. The nucleus is, therefore, the seat of matter, and in order to find out about the structure of matter it is necessary to tackle the problem of the structure of the nucleus.

The first clear concepts of the constitution of the nucleus emerged in 1919. By that time it had been established, mainly due to the work of F. W. Aston, that most elements are not simple, but consist of a mixture of several types of atoms differing from each other in weight. The various types of atoms of a given element are called isotopes; all isotopes of one element have the same atomic number, i.e. the same number of electrons in their atoms, and consequently they have all the same chemical properties, but their nuclei have different weights. The analysis of the isotopic composition of an element can be carried out by means of a method analogous to that used by J. J. Thomson to determine the properties of the electron. An instrument called the mass spectrometer is used, which employs a combination of electric and magnetic fields. The atoms of the given element are ionized by knocking out electrons from them and thus making them positively charged. In the electric and magnetic fields ions of different masses are deflected by different amounts. By observing into how many groups a given beam of ions splits, the number of isotopes can be determined; by measuring the deflection of each group the atomic weight of the individual isotopes can be calculated.

The Periodic Table of elements (main table, groups left to right):

1	2	3	4	5	6	7	8	9	10	11	12	13	14	15	16	17	18
1 Hydrogen H																	2 Helium He
3 Lithium Li	4 Beryllium Be											5 Boron B	6 Carbon C	7 Nitrogen N	8 Oxygen O	9 Fluorine F	10 Neon Ne
11 Sodium Na	12 Magnesium Mg											13 Aluminium Al	14 Silicon Si	15 Phosphorus P	16 Sulphur S	17 Chlorine Cl	18 Argon A
19 Potassium K	20 Calcium Ca	21 Scandium Sc	22 Titanium Ti	23 Vanadium V	24 Chromium Cr	25 Manganese Mn	26 Iron Fe	27 Cobalt Co	28 Nickel Ni	29 Copper Cu	30 Zinc Zn	31 Gallium Ga	32 Germanium Ge	33 Arsenic As	34 Selenium Se	35 Bromine Br	36 Krypton Kr
37 Rubidium Rb	38 Strontium Sr	39 Yttrium Y	40 Zirconium Zr	41 Niobium Nb	42 Molybdenum Mo	43 Technetium Tc	44 Ruthenium Ru	45 Rhodium Rh	46 Palladium Pd	47 Silver Ag	48 Cadmium Cd	49 Indium In	50 Tin Sn	51 Antimony Sb	52 Tellurium Te	53 Iodine I	54 Xenon Xe
55 Caesium Cs	56 Barium Ba	57 Lanthanum La	72 Hafnium Hf	73 Tantalum Ta	74 Wolfram W	75 Rhenium Re	76 Osmium Os	77 Iridium Ir	78 Platinum Pt	79 Gold Au	80 Mercury Hg	81 Thallium Tl	82 Lead Pb	83 Bismuth Bi	84 Polonium Po	85 Astatine At	86 Radon Rn
87 Francium Fr	88 Radium Ra	89 Actinium Ac	104 Rutherfordium Rf	105 Hahnium Hn													

Lanthanide series:

58 Cerium Ce	59 Praseodymium Pr	60 Neodymium Nd	61 Promethium Pm	62 Samarium Sm	63 Europium Eu	64 Gadolinium Gd	65 Terbium Tb	66 Dysprosium Dy	67 Holmium Ho	68 Erbium Er	69 Thulium Tm	70 Ytterbium Yb	71 Lutecium Lu

Actinide series:

90 Thorium Th	91 Protactinium Pa	92 Uranium U	93 Neptunium Np	94 Plutonium Pu	95 Americium Am	96 Curium Cm	97 Berkelium Bk	98 Californium Cf	99 Einsteinium Es	100 Fermium Fm	101 Mendelevium Md	102 Nobelium No	103 Lawrencium Lr

Fig. 1. The Periodic Table of elements

The determination of the atomic weights of individual isotopes revealed the amazing fact that they were all very nearly integral numbers, that is to say, that they could be expressed as almost exact multiples of the atomic weight of hydrogen. Thus, for example, chlorine, which has an atomic weight of 35·457, was found to be a mixture of two isotopes, one of which has an atomic weight of 34·978 and the other 36·977; similarly copper was found to have two isotopes of atomic weights nearly 63 and 65. This discovery brought back to life the old hypothesis of Prout in a slightly modified form; it has now become possible to assume that the nuclei of all elements are built up of the nucleus of the hydrogen atom. This nucleus, which carries one elementary positive charge, was given the name 'proton'.

Although this assumption was very attractive, it would have been little more than a hypothesis without direct proof that protons are indeed present in nuclei of other elements. This proof was given by Rutherford in an epoch-making experiment in 1919, in which he effected the disintegration of the nitrogen nucleus, the first man-made transmutation of elements. Rutherford bombarded nitrogen atoms with a-particles from radioactive substances and demonstrated that protons were emitted as a result of the disintegration. Later, similar experiments were carried out with other elements and in these cases too the emission of protons was established beyond doubt. Thus, definite evidence was obtained that protons enter into the structure of nuclei of other elements and that, therefore, the proton is an elementary particle of matter.

The Neutron

The disintegration experiments provided definite proof that protons are components of nuclei of all elements, but protons by themselves are not sufficient to make up a nucleus. This can be seen clearly when one considers the relationship between atomic number and atomic weight of any element. Oxygen, for example, occupies the eighth place in the Periodic Table. This means that the atom of oxygen contains 8 electrons revolving round its nucleus, and consequently the nucleus of oxygen must contain

8 positive charges, i.e. 8 protons. On the other hand, the atomic weight of oxygen is 16, which means that the nucleus is 16 times heavier than the hydrogen atom, and so would require the presence of 16 protons. How can we account for this discrepancy? The simplest way is to assume that in addition to the orbital electrons there are some electrons in the nucleus. Thus, the nucleus of oxygen might consist of 16 protons and 8 electrons. The 8 electrons would neutralize 8 positive charges of the protons, making an effective charge of 8, while the total mass would still be 16 since the electrons hardly contribute to the weight. In a similar way, the charge and mass of all other nuclei could be explained by assuming that apart from protons they contain a certain number of electrons. The existence of electrons in the nucleus is seemingly also supported by the fact that some of the radioactive elements emit electrons when they break up. It might be argued that if an electron can be emitted from a nucleus, it must have been there before.

In this way the whole problem of the structure of matter became greatly simplified. All matter could be assumed to be built up of only two particles, protons and electrons. This solution, though attractive in its simplicity, as it brought down the number of components of matter to a minimum, could not, however, be upheld, and a number of flaws were soon found in it. The main objection was the assumed presence of electrons in the nucleus. As the development of nuclear physics proceeded, more and more experimental and theoretical evidence accumulated, which was at variance with this simple scheme, and which particularly contradicted the idea that electrons were present in the nucleus. One of these difficulties can be explained by means of the following example of the nucleus of nitrogen. Nitrogen has atomic number 7 and atomic weight 14. We would, therefore, have to assume that its nucleus contains 14 protons and 7 electrons, altogether 21 particles. On the other hand, from the molecular spectrum of nitrogen, as well as from other evidence, it appears that the nucleus of nitrogen must contain an even number of particles.

Perhaps the strongest argument against the existence of electrons in the nucleus is of a theoretical nature. One can calculate the magnitude of the force needed to confine the negatively charged electron within the boundary of the positively charged nucleus, and this force turned out to be several hundred times greater than that actually observed in the nucleus. All these considerations add up to the simple statement that there is no place for electrons in the nucleus. But, if so, what is there to take the place of these electrons?

The answer to this question came in 1932, when James Chadwick discovered the neutron. The neutron is a particle of approximately the same mass as the proton, but without an electric charge. Its discovery was the result of experiments carried out by physicists in several countries. First, W. Bothe and H. Becker in Germany observed that when some light elements are bombarded with a-particles, they send out a very penetrating radiation which they thought to be γ-rays of a high energy. Next, Irene Curie and Frederick Joliot in France found that if these rays are made to collide with hydrogen atoms, fast protons are emitted. The French scientists still upheld the γ-ray hypothesis and they thought that they had discovered a new mode of interaction of γ-rays with matter. Chadwick, however, immediately recognized the true nature of the penetrating rays and produced experimental evidence that they were particles and not waves. He deduced that the mass of these particles was about the same as that of the proton, and gave the proper interpretation for their great penetrating power as due to the absence of electric charge, which means that they do not interact with electrons of the atoms through which they pass. Very soon afterwards it was found that a similar emission of neutrons is observed when other elements are bombarded with a-particles, and this provided proof that the neutron is a constituent of the nucleus.

Immediately after the discovery of the neutron it was recognized that it was no longer necessary to assume the presence of electrons in the nucleus. Thus, for example, in order to explain the charge and mass of the nucleus of oxygen, one need only assume that it contains 8 protons and 8 neutrons. The total mass is, therefore, 16, while the charge is only 8. Similarly, nitrogen would contain 7 protons and 7 neutrons, a total of 14 particles, an even number. Thus, electrons were no longer required to be present in the nucleus and the structure of all atoms could be explained as consisting of protons and neutrons in the nuclei and of electrons outside them.

This is the view still held now. All atoms are assumed to be built up of three elementary particles: protons, neutrons and electrons. The protons and neutrons make up the nucleus of the atom. They have the collective name 'nucleons', and the total number of nucleons is called the mass number of the isotope. This is so because in the conventional system of atomic weights the mass number is very nearly the same as the atomic weight. Different isotopes of the same element have the same number of protons in their atoms but different numbers of neutrons. The number of protons is equal to the number of electrons revolving round the nucleus and this number is the atomic number of the element. The constitution of every nuclear species, or nuclide, is, therefore, completely defined by two numbers, the atomic number and the mass number. It is usual to denote a nuclide by the symbol of the element, with the mass number as a superscript on the left side, e.g. ^{16}O, ^{23}Na, ^{226}Ra; the atomic number is sometimes given as a subscript on the left side ($^{16}_{8}O$, $^{23}_{11}Na$, $^{226}_{88}Ra$) but usually it is omitted, it being assumed that everyone knows the Periodic Table by heart!

The Positron

Although the problem of the constitution of atoms had thus been satisfactorily solved, there still remained the question whether the proton, neutron and electron are the only elementary particles. Almost before the simple theory outlined above was established, facts became known which demanded the existence of a few more.

The conclusion that electrons do not exist in the nucleus has probably raised a question in the reader's mind. How do we explain the phenomenon of β-decay, the spontaneous emission of electrons from the nuclei of radioactive atoms? If electrons do not exist in the nucleus, how do they suddenly appear at the moment when they are to be emitted? The answer is that this is exactly what happens. Electrons are not normally in the nucleus but are created there at the moment when they are to be ejected. This is a process analogous to the emission of light from the atom, or of heat from a burning substance. The light or the flame were not there before but were produced at the required instant.

At first sight this analogy may not appear to be justified, because in the case of light or heat we deal with energy, while in the case of β-decay we deal with the creation of a particle, a fragment of matter. This objection is, however, removed when one takes into account the principle of the equivalence of matter and energy, as formulated by Einstein in his theory of relativity, which will be discussed in Chapter 6. In accordance with this, matter and energy are equivalent; matter can change into energy and vice versa. Since the electron has a very small mass, a relatively small amount of energy, which can be readily found in the nucleus, is sufficient to create it. Thus, the emission of an electron from the nucleus does not imply the production of matter from nothing, but only a transformation from energy.

The electron, however, apart from having a mass has also an electric charge and this too cannot be created from nothing. The law of conservation of electric charge requires that a positive charge is created at the same time. In the case of β-decay this is taken care of by the transformation of a neutron into a proton, as will be described in the next section. However, there can be a pure transformation of energy into matter, the so-called materialization process, and in this case the creation of an electron must be accompanied by the creation of a particle with a positive charge, a positive electron.

The possible existence of a positive electron, or positron, was first foreshadowed in 1928, when Dirac formulated his theory of the electron, from which it followed that apart from the negative electron there must also exist its positive counterpart. But it was not until 1932 that the positron was discovered experimentally. C. D. Anderson, an American scientist, studied the cosmic radiation by means of a cloud chamber. In this instrument the path of a charged particle can be rendered visible by suddenly expanding a volume saturated with water vapour. Droplets of water are then deposited on the ions formed along the path of the particle, which thus produces a trail somewhat similar to that left by an aircraft. If the cloud chamber is placed in a magnetic field, the path of the particle is curved and from the direction of curvature the sign of the charge can be deduced; from the density of the tracks it is possible to discriminate between a proton and an electron. The photograph given in Plate I is of historic interest as it shows the first track identified as that of a positron. The cloud chamber contained a metal plate and it is seen that the curvature of the track is greater above the plate than below. Since the curvature is the greater the smaller the energy, and since the particle must have lost some energy in passing through the plate, this proves that the particle must have been moving upwards. From the observed curvature and from the known direction of the magnetic field Anderson deduced that the particle must have been positively charged.

Soon after this discovery was made it turned out that it is not necessary to look for positrons in cosmic radiation because they can be very easily produced by several methods available in the laboratory. One of these is the already mentioned process of materialization of energy, i.e. the creation of a negative and positive electron from energy. Such pair production, as it is called, can be brought about when γ-rays of sufficiently high energy pass through the field of an atom. The γ-ray may then disappear altogether and in its place may appear a pair of particles, an electron and a positron. Plate II shows such a process of pair production occurring in a cloud chamber. A beam of γ-rays (which do not show up in the cloud chamber because they are uncharged) is passed through the chamber and at the point indicated by the arrow a γ-ray has struck an atom of the gas filling the chamber and produced a pair of charged particles. Both tracks are of the same density but they are curved in opposite directions which shows that they are of opposite charges. Another source of positrons is the spontaneous emission of positive electrons from some radioactive substances, a process similar to β-decay. Thus, there seems to be no doubt about the existence of the positive electron.

If the positron is an elementary particle the question may be raised: why does not the positron occur in nature as frequently as the electron? The reason is that soon after a positron is

13

created it disappears as a result of a collision with an electron. This process is the reverse of pair production, annihilation of matter as contrasted with materialization of energy. When two opposite electrons meet they combine together and are converted into energy, into two γ-rays, or much less frequently into three γ-rays. Since our universe contains so many negative electrons, it follows that very soon after a positron is born it meets an electron and is annihilated. The average time the positron exists before annihilation is very short, about 10^{-12} seconds, but despite its short existence its properties have been fully investigated and found to be the same as those of the electron, except for the sign of the charge.

The Neutrino

We must now return to the problem of β-decay and explain how an electron, or a positron, is produced in a nucleus at the instant when it is to be emitted. This touches on one of the most difficult subjects in nuclear physics, the problem of nuclear forces. The components of nuclei, protons and neutrons, act on each other, that is protons on neutrons, protons on protons and neutrons on neutrons, with certain specific attractive forces, which are quite unlike the electrical or gravitational forces with which we are familiar from everyday life. From various data it can be deduced that these nuclear forces are unlike electrical or gravitational forces in that they are very short-range forces, they act only when the two particles are very close together inside the nucleus; if the particles are a little bit further apart the force is no longer there (on the other hand, at still much shorter distances these forces must change from attractive into repulsive, otherwise the nucleons would coalesce). There is also evidence that the nuclear forces are of an exchange type, which means that the particles attract each other by virtue of a feature which they keep exchanging. Such a feature is their identity; for example, the proton can change into a neutron and the neutron into a proton. This action was initially visualized as follows: a neutron may be considered to be a combination of a proton and an electron. When a neutron collides with a proton it gives off the electron, turning itself into a proton; the electron is taken up by the proton, which becomes a neutron. In this way the two particles have exchanged identities and it is this continuous exchange which constitutes the attraction between them. Similarly, a proton may be considered to be a neutron plus a positron, and the latter is exchanged with a neutron encountered. One may deduce from this that the proton and neutron can be considered as two different states of the same particle; in one state it is a proton and in the other a neutron. The exchange of neutrons and protons was confirmed experimentally from observations on collisions of fast neutrons with hydrogen.

The process of exchange usually goes on in the nucleus without producing any change in its constitution or emission of any radiation. There may, however, be circumstances when a nucleon changes its identity without an exchange with another. This happens in the case of an unstable or radioactive nucleus. A radioactive nucleus is one which has the wrong constitution, it has either too many neutrons or too many protons. In the first case this can be remedied if one of the neutrons becomes a proton. The electron which has to be created at the same time to preserve the balance of charge cannot exist in the nucleus and is immediately emitted. We observe then the emission of β-rays. Similarly, if a nucleus has a surplus of protons, one of the protons may change into a neutron. A positive electron is then created and immediately emitted. This explains the emission of positive electrons from some radioactive elements.

It may be remarked here that the neutron itself is an unstable particle. If it is allowed to exist outside the nucleus for a long enough time, it breaks up spontaneously into a proton and an electron in a matter of minutes; inside the nucleus, however, it can exist indefinitely by virtue of the exchange process.

The above scheme explains how electrons can be emitted from nuclei, but the process of β-decay presented yet other difficult problems. One of them concerns the energy of the emitted electrons. When radioactive atoms break up with the emission of β-rays the electrons are fired away with a considerable energy, which may be measured in a variety of ways, for example, by observing their deflection in a magnetic field. In accordance with the general ideas of quantum theory, which we discuss in Chapter

15

6, only certain well-defined values of energy can occur in nuclei. Indeed, α-particles and γ-rays are always emitted from a given nuclide with well-defined, discrete energy values; the same might, therefore, have been expected in the case of β-rays. Instead of this, it was found that electrons emitted from the nuclei of a given substance have a continuous distribution of energies, from zero up to a certain maximum value. The graph in Fig. 2 shows a

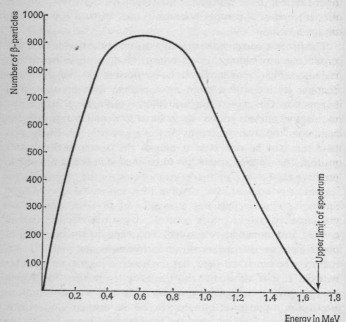

Fig. 2. Energy spectrum of β-rays

typical energy distribution of β-rays; the horizontal scale gives the energy, and the vertical scale the number of electrons which are emitted with that energy. It is seen that there is a continuous distribution in which all energies are represented, instead of only certain discrete values. The problem is, however, not only of the observation of a continuous instead of a discrete spectrum, but also of an unaccountable loss of energy. It is often possible to calculate from other data the amount of energy which has been released during the β-transformation, and which should have appeared as the energy of the emitted β-rays. Actually, in every case where such calculation is made, the total energy released is found to be equal to the upper limit of the spectrum, and as seen from the graph of Fig. 2 the electrons are emitted with less energy than this value. The question then arises: what has happened to the remainder of the energy? The cardinal law of physics, the law of conservation of energy, appears to have broken down in the case of β-decay.

Another law which appears to have broken down is that of the conservation of angular momentum, or spin. All elementary particles which we described so far behave as if they had a spinning motion with a well-defined angular momentum, which is given the value of $\frac{1}{2}$ in terms of certain constants. In all nuclear reactions the total value of the angular momentum remains unchanged. However, in the process of β-decay, in which one particle (say a proton) is transformed into two (a neutron and positron) each with a spin $\frac{1}{2}$, the angular momentum is obviously not conserved.

In order to save the two principal laws of physics a new particle had to be invented. This was done by Pauli, an Austrian physicist who lived in Switzerland, who proposed the neutrino hypothesis in 1931. According to this hypothesis, when a neutron changes into a proton, or vice versa, in addition to the electron yet another particle is created, the neutrino. This particle has no mass and no electric charge. For these reasons it does not interact with matter through which it passes and consequently it is practically impossible to observe it. Nevertheless, it performs a very important function: it takes away part of the energy available in the β-decay, the total energy of the decay being shared between the electron and the neutrino. In this way the continuous energy spectrum of the β-rays can be explained. Similarly, by assigning to the neutrino the same spin, $\frac{1}{2}$, as the other elementary particles have, the total angular momentum can be conserved.

Although the neutrino was originally invented only to preserve the conservation laws, it turned out to be of very great theoretical value. By using the neutrino hypothesis, Fermi was able to put

17

forward in 1934 a complete theory of β-decay, which explained quantitatively both the shape of the spectrum and the rate of decay.

Direct evidence of the existence of the neutrino came much later, in 1956, in an experiment which involved the use of a large nuclear reactor in the U.S.A., originally built for military purposes. The aim of the experiment was to detect the neutrino by an inverse process to that of β-decay: the neutrino is captured by a proton which is then transformed into a neutron and a positive electron. Both these products can be detected by means of an elaborate and very expensive apparatus designed to eliminate extraneous effects. Owing to the extremely weak interaction of neutrinos with matter, a very intense neutrino flux is essential; this became available only after the development of nuclear reactors (see Chapter 4) in which a large number of β-decay processes take place among the fission fragments. Despite the complexity of the experiment it has established beyond any doubt that the neutrino does exist. Later experiments have made it possible to establish that in fact two types of neutrinos exist (see next section).

Muons and Pions

In the previous section it was stated that the nuclear forces cause neutrons to change into protons and vice versa, and that in the case of unstable nuclides this transformation gives rise to the emission of electrons and neutrinos. On the basis of Fermi's theory of β-decay involving the exchange of two particles it is possible to calculate the strength of the interaction in such transitions. On the other hand, it is possible to measure the nuclear forces directly, from observations on collisions between protons and neutrons. A comparison of these two procedures has revealed an enormous gap. The forces calculated on the basis of Fermi's theory of β-decay were found to be extremely small compared with the nuclear forces deduced from collision experiments. Thus another serious conflict emerged between theory and experiment.

In an attempt to remove this difficulty the Japanese physicist

Yukawa put forward in 1935 the hypothesis that one particle is exchanged in the interaction between nucleons, and that this is not the electron but a new particle, which according to his calculations would have a mass several hundred times greater than that of the electron. Because of its mass being intermediate between that of the proton and electron, this hypothetical particle was given the name meson. Its main virtue is that it is a much stronger 'glue' than the electron to keep the nucleons together, and Yukawa has shown that the postulation of its existence will account for the very strong interaction as compared with processes, like β-decay, in which electrons and neutrinos are involved.

Yukawa's hypothesis did not receive initially much attention, but several years later a particle with apparent properties of the meson predicted by him was indeed discovered experimentally. As in the case of the positron a few years earlier, the new particle, which is now known as the muon, was discovered in cosmic radiation by a group of American scientists. They observed tracks of cosmic rays in the cloud chamber which could only be made by particles of unit charge and of mass about two hundred times the electron mass. Both positive and negative muons were observed, and it was also found that they were not stable but broke up spontaneously within a few millionths of a second, changing into electrons and neutrinos.

After this discovery it was thought that the muons from the cosmic radiation were just the particles needed to account for the magnitude of the nuclear forces. Soon, however, it became clear that this was not so. It was found that muons very rarely interacted with nuclei; they could pass through large thicknesses of matter with hardly any absorption. If so, they could not be expected to play an important role in binding the nucleons together. The old problem of the nuclear forces still remained unexplained.

It was not until 1947 that a satisfactory solution was found. In that year Powell and his collaborators at Bristol discovered another particle which they called π-meson, or pion. They employed the photographic emulsion technique which makes use of the fact that the passage of a charged particle through an emulsion renders the grains along its path developable; after processing the emulsion the path is revealed as a row of black grains which can be seen through a microscope. From the length of the track and the density of grains in it the mass and energy of the particle can often be determined. Powell exposed photographic plates to cosmic rays at high altitudes and found in them a number of events which could leave no doubt that they were due to the passage of particles of intermediate mass. It was also established that these particles decay after a short time and are transformed into the muons previously observed in cosmic radiations. The photograph of Plate III, which was obtained by Powell later with more sensitive emulsions, shows the successive transformations. The thick track on the left is that of a pion; it is seen that the density of the grains increases from bottom to top, indicating that the particle was moving in this direction. The π-meson then came to rest and gave rise to a muon, which produced its own track until it came to rest and in turn decayed into an electron. Further investigations with π-mesons have shown that they interact very strongly with nuclei and indeed behave like the particles predicted by Yukawa. It must, therefore, be assumed that it is the π-meson which keeps the nucleons together in the nucleus.

The initial investigations on pions had to be carried out with the small number present in the cosmic radiations, as no other source of mesons was known. Later, however, it became possible to produce mesons artificially in the large accelerating machines. If a beam of, say, protons of a very high energy is made to strike a light element, a copious emission of mesons is observed. In fact, the main object of the colossal accelerators which have been developed in recent years (see Chapter 3) was to produce mesons, as well as the other particles discovered subsequently. By these means it was possible to obtain intense beams of π-mesons, and this has greatly facilitated the study of their properties.

We know now that there are three types of pions, positive, negative and neutral. The mass of the charged pion is 273 electron masses, while that of the neutral π-meson is 264 electron masses. Unlike other elementary particles pions have no intrinsic angular momentum (spin = 0). All π-mesons are very short-lived; the mean life of the charged pion is about 10^{-8} second, while that of the neutral pion is much shorter, about 10^{-16} second. At the end of their life charged π-mesons decay into muons and neutrinos; a neutral π-meson decays into two γ-rays.

The big divergence between the magnitude of the force involved in nucleon interactions and that in β-decay has necessitated the postulation of two types of nuclear forces, which are now called strong interactions and weak interactions. The strong interactions occur between nucleons and involve pions: a proton will emit a positive pion and become a neutron, while another neutron will absorb the positive pion and become a proton; similarly, a neutron can become a proton by the emission of a negative pion which is then absorbed by another proton; finally, the forces between like nucleons, e.g. between a proton and a proton, or between a neutron and a neutron – which were found to be of the same magnitude as between protons and neutrons – can be explained as due to the exchange of neutral pions. The weak interaction operates in β-decay, when a neutron changes into a proton (or vice versa) with the emission of an electron and neutrino.

The two types of interaction differ enormously in magnitude; the strong interaction is very much stronger than electrical forces, which in turn are stronger than the weak interactions. Apart from the strength, there are also other fundamental differences between the two types of interaction, relating to the conservation of certain symmetries.

The transition of a pion into a muon plus neutrino belongs to the weak interactions, but it has been shown that the neutrino emitted in this process is different from the neutrino emitted in β-decay. Apart from the fact that the muon mass is 207 times greater than that of the electron, the muon appears to have the same properties as the electron and is sometimes referred to as a heavy electron. Muons, electrons and neutrinos are usually classified together under the collective name of 'leptons' (light particles).

Strange Particles

With the exception of muons, whose role in nature is still not clear, all the other elementary particles discussed so far fit in very well into an overall picture of the structure of matter and are sufficient to explain its basic properties. One might, therefore, have thought that with the discovery of π-mesons the list of elementary particles was complete. But this was not to be the case. Almost simultaneously with the discovery of the π-meson, evidence of yet another, still heavier meson, called K-meson, or kaon, appeared. At about the same time other particles, heavier even than the nucleons, and called hyperons, began to be discovered; since 1948 a large number of these strange particles were found.

The discovery of the relatively stable of the new particles followed the same pattern as the discovery of the positive electron, muon or pion. First they are observed in cosmic radiation, usually by scanning photographic emulsions which have been sent up to the upper regions of the atmosphere in balloons, and later in rockets and space probes. Since these particles are not very common, huge areas of emulsion have to be scanned in order to find enough events to establish the main properties of the new particle: mass, charge, mean life, mode of decay, etc. Later the accelerator physicists take over; by achieving a sufficiently high bombarding energy of the projectiles from the accelerator, they are able to produce the new particles in the laboratory. Usually copious beams of such particles can be produced and the study of their properties then proceeds much more quickly. Since the heavier the particle the larger the energy needed to produce it, there is usually a lag of several years between the discovery of a particle in cosmic radiation and the building of an accelerator of sufficiently high energy to produce it.

The study of new particles produced in accelerators has been made easier by the development of a new instrument, the bubble chamber. This is based on the same principle as the cloud chamber but instead of a gas a super-heated liquid is used. A pure liquid can normally be heated to a temperature higher than the boiling point without boiling, but if an ionizing particle is passed through the liquid it triggers local boiling along the path of the particle, forming bubbles which make a visible track. Owing to the much higher density of liquids as compared with gases, various processes occurring during the passage of particles of very high energy can be observed in a bubble chamber under conditions in which a cloud chamber would hardly register anything. Plate IV shows a bubble-chamber photograph of a K-meson and a hyperon produced in a collision between a pion and a proton.

Various types of K-mesons, positive, negative and several variations of neutral, with masses nearly 1,000 times the electron mass, have been identified. Several types of hyperons, with different masses, lifetimes and modes of decay have also been discovered at about the same time. The hyperons were given capital Greek letters: Λ, Σ, Ξ, Ω. The hyperon on Plate IV is a Σ^-; it subsequently decays into a pion and a neutron, the latter not leaving a visible track.

During the following years, as larger accelerators came into being, and intense beams of very high energy protons and pions became available, the number of these strange particles increased rapidly, and by 1972 some hundreds of these mesons and hyperons were tabulated. Almost all of them were characterized by extremely short lifetimes, of the order of 10^{-24} seconds, and this gave the clue to their nature. It became obvious that they were not new particles, but nucleons and hyperons excited to high energy states. Just as atoms or nuclei can exist for a very short time in excited states, if they are provided with sufficient energy, so the nucleon itself can have a number of excited states, or resonances as they are called. A system in an excited state is unstable, and the transition to a lower state, and eventually to the lowest, or ground state, usually occurs in a very short time and is accompanied by the emission of energy. In the case of the atom this energy appears in the form of light, the familiar atomic spectra. In the case of nuclei it appears either as γ-rays, which are the same electromagnetic waves as light but of shorter wavelengths, or in the form of β-decay. Both these processes can also occur in nucleon resonances, but a third process occurs here most frequently, the emission of a meson, a π, K, or other variety. Thus, mesons play a similar role in the nucleons, as β-spectra in nuclei and light spectra in atoms. The observation of the numerous resonances and mesons can thus be explained without the need to postulate the existence of new particles.

Very soon after the discovery of the resonances some regularities began to be observed in their mode of decay. Gell-Mann and others suggested in 1961 that the different resonances can be arranged in a few groups, each of which is characterized by an integral number, called 'strangeness'. The nucleons have the strangeness number 0, while for the 'new particles' it can be 1 or 2 or 3 units. Just like the conservation of mass, charge, spin, or number of nucleons, which applies to the familiar particles, there is also a conservation of strangeness, so that only certain transitions between resonances are allowed. This rule has helped to bring order to the apparent chaos of the resonances.

Within each strangeness group there are one or two states which differ from all the others by having much longer lifetimes than the other resonances. On an absolute scale these lifetimes are still extremely short, 10^{-10} or 10^{-14} seconds, but on a nucleon scale they are very long, billions of times longer than the other resonances. By comparison they are therefore almost stable, and for this reason they are usually classified as particles. These are the Λ-, Σ-, Ξ- and Ω-hyperons mentioned earlier. The Λ and Σ particles have strangeness of one unit, the Ξ type has strangeness 2 and the Ω type 3 units.

The reason for the relatively long lifetimes is that the ordinary transition to the ground state via mesons is forbidden by certain selection rules, and they can only decay by weak interactions involving leptons. It should be noted that such 'metastable' states are also known in atomic and nuclear transitions. Gell-Mann's theory of strangeness numbers is based on the so-called SU(3) model, which stands for symmetric unitary groups for arrays of size 3×3. It is a generalization of the theory which explains the different states observed in nuclei. The success of the SU(3) model was assured when it successfully predicted the existence of the Ω-hyperon, and described its properties long before it was discovered in 1964.

Anti-Particles

Earlier in this chapter it was stated that the existence of the positive electron was theoretically predicted by Dirac. Actually, his theory applies not only to the electron, but to any elementary particle. Dirac attempted to find the equation of motion of a free particle of spin $\frac{1}{2}$ in terms of quantum mechanics and the theory of relativity, and to his surprise he found that the equation gave two symmetrical solutions. In practical terms this means that to every particle there must exist its counterpart, an anti-particle, which has the same mass and other properties but a charge of opposite sign. Thus, a positive particle will have a negative anti-particle. Neutral particles, e.g. the neutron, neutrino, neutral mesons or hyperons, also have anti-particles which spin in the opposite direction to that of the particles.

Dirac's theory not only predicted the existence of anti-particles but also the conditions for their production and their subsequent fate. An anti-particle can be produced together with its particle in any nuclear collision in which a sufficiently large amount of energy is available. This is the process of pair production mentioned earlier. In accordance with the principle of equivalence of mass and energy, the heavier the particle, the larger the amount of energy necessary to produce it. After having been produced, the anti-particle – although intrinsically stable – has only an ephemeral existence; as soon as it collides with its particle both of them undergo annihilation, their masses being converted into energy.

These predictions, earlier confirmed in the case of the electron–positron pair, met their real test in the discovery of the anti-proton. On the basis of Dirac's theory, it was to be expected that the counterpart of the proton, a proton with a negative charge, should exist, but it was not until 1955 that a sufficiently high energy was achieved in an accelerator to produce it. In the proton-synchrotron at Berkeley a beam of protons of very high energy was made to impinge upon a copper target. Various types of particles were produced in this collision, but by means of an ingenious apparatus it was possible to demonstrate that one of these had a mass equal to that of the proton but negative charge; it had in fact all the properties of the anti-proton. Furthermore, it was found that this particle very soon undergoes annihilation at a collision with another proton. Plate V shows a bubble-chamber photograph of the annihilation of an anti-proton; a number of π-mesons are produced which eventually decay into electrons, neutrinos and γ-rays. Many such events have been observed, so that the existence of the anti-proton and its properties are established beyond doubt.

Owing to the exchange nature of the proton and neutron, an anti-proton may be expected occasionally to change into an anti-neutron. Beams of such anti-neutrons have indeed been observed and their properties investigated. Anti-particles of the hyperons have also been observed.

The discovery of anti-protons has given rise to speculations about the possible existence of anti-matter, that is atoms which consist of anti-protons, with positive electrons revolving around them. Such matter would have exactly the same properties as ordinary matter, but if the two types of matter collided all of it would undergo annihilation with a huge release of energy. It has been suggested that some galaxies in the remote parts of the universe are made up of such anti-matter, and that the occasional violent explosions which one observes in these galaxies are the result of collisions between matter and anti-matter. For the time being, these interesting suggestions must remain in the realm of speculation.

Quarks – the Ultimate Particles?

Our present-day knowledge of elementary particles, stable and metastable, is summarized in Table 1 in which the particles are listed in order of increasing mass. Since all the charged particles carry one unit of electrical charge only the sign is given. The spin, strangeness number and mean life are also listed. For hyperons, each of those listed has an anti-particle with a positive strangeness. All masses are given in units of the electron mass. For sake of completeness, the photon, the quantum of electromagnetic radiation energy, is also included. The first column of the Table gives the collective names by which groups of particles are known.

It will be seen from Table 1 that, apart from the leptons, the only really stable particle is the proton. But can we consider the proton to be the ultimate particle, or could it in turn be broken

Table 1. Elementary Particles

Group	Name	Symbol and charge	Mass	Spin	Strange-ness	Mean life (seconds)
	Photon	γ	0	1	0	Stable
Leptons	e-neutrino	ν_e	0	$\frac{1}{2}$	0	Stable
	mu-neutrino	ν_μ	0	$\frac{1}{2}$	0	Stable
	electron	$e^\pm$	1	$\frac{1}{2}$	0	Stable
	muon	$\mu^\pm$	206·8	$\frac{1}{2}$	0	$2·2 \times 10^{-6}$
Mesons	Neutral pion	π^0	264·1	0	0	$0·8 \times 10^{-16}$
	charged pion	$\pi^\pm$	273·1	0	0	$2·6 \times 10^{-8}$
	charged Kaon	$K^\pm$	966·4	0	± 1	$1·2 \times 10^{-8}$
	neutral Kaon	K^0, anti K^0	974·1	0	± 1	$0·9 \times 10^{-10}$
Baryons	Nucleons:					
	protons	$p^\pm$	1 836·1	$\frac{1}{2}$	0	Stable
	neutrons	n, anti n	1 838·6	$\frac{1}{2}$	0	$0·93 \times 10^3$
	Hyperons:					
	Lambda	Λ^0	2 183·1	$\frac{1}{2}$	-1	$2·5 \times 10^{-10}$
	Sigma	Σ^+	2 327·6	$\frac{1}{2}$	-1	$0·8 \times 10^{-10}$
	,,	Σ^0	2 333·6	$\frac{1}{2}$	-1	$<1·0 \times 10^{-14}$
	,,	Σ^-	2 343·1	$\frac{1}{2}$	-1	$1·5 \times 10^{-10}$
	Xi	Ξ^0	2 572·8	$\frac{1}{2}$	-2	$3·0 \times 10^{-10}$
	,,	Ξ^-	2 585·6	$\frac{1}{2}$	-2	$1·7 \times 10^{-10}$
	Omega	Ω^-	3 273·0	$\frac{3}{2}$	-3	$1·3 \times 10^{-10}$

down into other components? The fact that the nucleon can exist in a number of excited states may be taken as an indication that it has some structure. Just as from observations on atomic spectra the structure of the atom was deduced, so the regularities observed in nucleon resonances and the meson and hyperon spectra may give a clue to the structure of the nucleon.

In recent years a new hypothesis, with some curious assumptions, has been put forward to explain the various phenomena observed with the mesons and baryons. This theory postulates the existence of a new hypothetical particle called 'quark'. Actually, in the light of the three dimensions of the SU(3) model, three types of quarks (with their corresponding anti-particles) are postulated; they are called p-, n- and λ-quarks. The first two have strangeness 0, and the third -1. All three have spins $\frac{1}{2}$ units. The most curious assumption is in relation to their electric charge. Despite the fact that all known charged particles carry a charge which is either equal to or a multiple of the electron charge, and that a smaller charge has never been observed, for the quarks a fractional charge is assumed: $+\frac{2}{3}$ for the p-quark and $-\frac{1}{3}$ for the n- and λ-quarks. The charges of the anti-quarks are the opposite to those of their quarks.

According to this model, the proton is made up of one n-quark and two p-quarks, giving it thus one positive unit of charge and a spin $\frac{1}{2}$ (the spins of two of the quarks cancel each other). The neutron is made up of one p-quark and two n-quarks, giving a total charge zero. The Λ^0-hyperon is made up of one quark of each type, giving it charge zero and strangeness -1. The Ω^--hyperon consists of three λ-quarks. In this way the properties of all baryons can be worked out.

With regard to mesons they are made up of two quarks, of which one is an anti-quark. For example, the π^+-meson consists of one p-quark and one n-anti-quark. This gives a total charge of one unit (the charge of the n-anti-quark is $+\frac{1}{3}$), zero spin, and zero strangeness, as observed. The K^0-meson is made up of one n-quark and one λ-anti-quark, resulting in total charge and spin zero, and strangeness 1.

The fact that the observed properties of so many mesons and baryons can be explained by means of only three particles gives considerable strength to the hypothesis, but so far, and despite intensive search, the quark remains a hypothetical particle. There were reports of the detection in cosmic radiation of particles with the properties of the quarks, but these have not been confirmed. Anyhow, even if their existence is definitely established it will still leave open the question of the nature of the forces acting between the quarks, or whether they are the ultimate particles. The energies involved in interactions between baryons and mesons are a thousand million times greater than those within the atom, but from studies of cosmic radiation it is known that energies a thousand million times greater still may be involved in exploding galaxies. We are still a long way from understanding the ultimate structure of matter.

CHAPTER 2

The Atom Breaks Up

The Radiations from Radioactive Elements

As already mentioned in Chapter 1, soon after the discovery in 1896 of radioactivity, i.e. the spontaneous emission of radiations from uranium and several other heavy elements, it was observed that these radiations consisted of three distinct types, which for brevity and convenience were termed α-, β- and γ-rays. The analysis of the radiations into the various groups can best be effected by using a magnetic field. In such a field charged particles are deflected, positive particles in one direction and negative in the opposite direction, while uncharged radiation remains unaffected.

Fig. 3 shows the result which would be obtained if a strong magnet were brought near a radioactive substance in such a way that the north pole was above the plane of the paper and the south pole beneath it. It is seen that one group of rays is deflected slightly to the left, which means that these must be particles carrying a positive electric charge. This group is called 'α-particles'. The fact that they are all deflected uniformly indicates that they are all emitted with the same energy. Another group of rays is seen to be much more strongly deflected to the right, which indicates that they are negatively charged particles. They are spread out widely, which proves that the particles are emitted with various energies. This group was given the name 'β-rays' and, as we know already, they are fast electrons.

Finally, the third group which passes through the field undeviated are called 'γ-rays'. The fact that they are unaffected by the magnetic field may mean either that they are particles with no electric charge or a pure wave, an electromagnetic radiation.

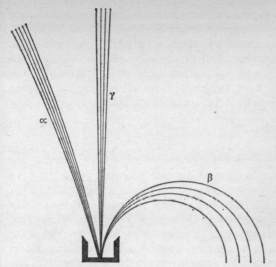

Fig. 3. Deflection of α-, β- and γ-rays by a magnetic field

Later it became possible to demonstrate directly that γ-rays show the phenomena of interference and diffraction, much in the same way as light waves do. This proves that they are electromagnetic waves, of the same nature as X-rays. In fact, there is no difference between X-rays and γ-rays. Initially the distinction between them was that γ-rays were of much shorter wavelength than X-rays, but nowadays we can produce X-rays of similar wavelength and indeed much shorter than those of γ-rays emitted from radioactive substances. The difference in nomenclature has, however, been preserved to indicate their origin: we call X-rays those originating from outside the nucleus, while by γ-rays we mean rays coming from the nucleus.

γ-rays may be considered as a secondary by-product of radioactive decay. The emission of an α- or β-particle results in a transformation of the radioactive element. Very often the nucleus produced as a result of this transformation is formed in an excited state, which means that it has more energy than the normal nucleus would have. This surplus energy is then got rid of in the form of electromagnetic radiation, which in this case we call γ-rays. It is a process similar to that occurring in the atom when it is in an excited state; the return to the normal state is accompanied by the emission of radiation in the form of light. Since the energies contained in the nucleus are about a million times greater than those involving the outer shell of the atom, the γ-rays emitted from radioactive substances have an average energy about a million times greater than the energy carried by light waves.

The α-particle

From the point of view of the role played in elucidating the structure of the nucleus the most important turned out to be the α-particles. It was Rutherford who very early and intuitively recognized the α-particle as the most powerful tool for probing the structure of the atom, and most of his work was concerned with this particle. Using methods similar to those described in the case of electrons, i.e. deflection in electric and magnetic fields, Rutherford was able to measure the velocity of the α-particles and the ratio of their charge to mass. In other experiments, in which the total charge carried by a known number of α-particles was measured, he was able to determine the charge of each α-particle.

These experiments proved that all α-particles emitted from radioactive substances are identical in nature: their charge is positive and twice the charge carried by an electron, and their mass is four times that of the hydrogen atom. According to our present-day views we realize immediately that the α-particle must be the nucleus of the helium atom, and composed of two protons and two neutrons. In those days, however, nothing was known about the nucleus, and the conclusion that an atom of one chemical element, helium, is emitted from the atom of another chemical element, say, radium, seemed so revolutionary that more direct evidence about its nature was necessary before it would be accepted. This piece of evidence was provided by Rutherford in the following experiment (Fig. 4). A quantity of radon, a radioactive gas, was compressed into the tube T which had walls thin enough to allow the α-particles emitted from radon to penetrate into the outer tube O. This latter tube was very thoroughly evacuated at the beginning of the experiment when a spectroscopic analysis showed no traces of helium in it. After leaving the radon for several days it was found that some

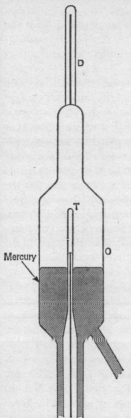

Fig. 4. Experiment to prove identity of α-particles

gas accumulated in the outer tube; this gas was then compressed, by bringing up the level of the mercury, into the discharge tube D, where its content could be analysed by observing the spectrum

33

of light emitted after an electric discharge was produced. The complete spectrum of helium was observed, and thus direct proof was obtained both of the identity of a-particles with helium and of the production of helium from radon.

As already mentioned, the a-particles from a given radioactive nuclide are all emitted with the same energy. This can be verified by deflecting the beam of a-particles in a magnetic field and measuring their energies directly, or more simply, although indirectly, by putting a source of a-particles into a cloud chamber and taking a photograph of their tracks. Such a photograph is shown in Plate VI. Each of the white lines represents a track of an a-particle. It is seen that the tracks are almost all straight lines, which indicates that the a-particle can generally pass through a thickness of air without changing its direction. The tracks end abruptly, and the distance travelled by an a-particle is called its range. In the photograph two definite ranges are seen; in air at atmospheric pressure the range of one group would be 4·8 cm and of the other 8·6 cm. There are two ranges because the source contained in this case two radioactive nuclides ^{212}Bi and ^{212}Po, which used to be called thorium-C and thorium-C'; each of these emits a-particles of a definite range. The range depends on the energy; the higher the energy of the particle the further it can penetrate. From the measurement of the range it is possible to determine the energy of the particle.

In nuclear physics energies are measured in electron-volts rather than in the usual units, joules. If a proton or an electron is accelerated in an electric field it acquires an energy which is proportional to the voltage across the field. We define the electron-volt as the amount of energy gained by an electron or a proton in passing through a potential difference of one volt. An a-particle, being double charged, will acquire double the energy in passing through the same difference of potential. In these units the energies of the two groups of a-particles from thorium-C and C' are 6·1 and 8·8 million electron-volts (MeV). Among the various natural radioactive elements the a-particles from thorium-C' have the highest energy; the lowest, 4·0 MeV, are from thorium-232.

Radioactive Transformations

Having established that radioactive substances emit spontaneously either helium nuclei or electrons, we may ask, what is the result of such emission? Rutherford and Soddy were the first to investigate this problem, but we shall consider it in the light of our present knowledge of the structure of the nucleus. We shall start with radium, which decays with the emission of an α-particle. This nuclide has an atomic number 88 and mass number 226, which means that it is composed of 88 protons and 138 neutrons. If an α-particle is emitted from it, i.e. 2 protons and 2 neutrons, the remaining nuclide will have atomic number 86 and mass number 222. In the Periodic Table (Fig. 1) element number 86 belongs to the group of inert gases (helium, neon, argon, krypton and xenon) and is called radon. Thus, as a result of the emission of an α-particle radium has changed into radon; we have here an example of transmutation of elements. This process continues with radon, which also emits an α-particle. The result of this decay is the formation of an element called polonium of atomic number 84 and mass number 218. Before the existence of isotopes was established this particular substance was given the name radium-A. Radium-A again decays by α-particle emission, producing a nuclide of atomic number 82, i.e. lead, and mass number 214.

This isotope of lead of mass 214 was called radium-B. Radium-B is found to decay by β-emission. As we know already, the emission of an electron is due to the transformation of a neutron into a proton. Since there has been no change in the number of nucleons, the nuclide produced in the transformation will have the same mass number 214, but since one of the neutrons has changed into a proton the charge of the nucleus has increased by one and consequently the atomic number of the product will be 83, which is that of bismuth. We find thus that the product of decay of radium-B is an isotope of bismuth which was called radium-C.

In a similar fashion we can follow the successive transformations of all the radioactive substances. Each time an α-particle is emitted the atomic number is decreased by two and the mass number by four; at each β-emission the mass number remains unchanged and the atomic number is increased by one. All radioactive substances found in nature among the heaviest elements can in this way be arranged into family trees, and it has been found that they form three separate radioactive series.

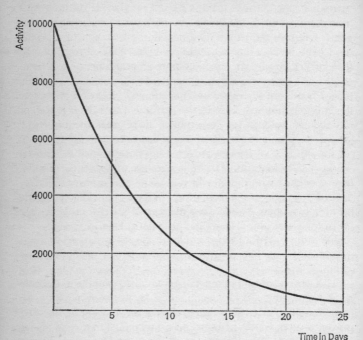

Fig. 5. Decay of activity of radioactive substance

Apart from its atomic number and mass number each radioactive substance is characterized by the rate at which its nuclei break up. In a given radioactive substance not all of its nuclei break up at the same time; some decay after a very short time, others may exist for a long time. It is a process governed by chance, but if the substance contains a very large number of unstable atoms a simple statistical law is observed, namely that in equal intervals of time equal fractions of all atoms present undergo disintegration. As a measure of the rate of decay one usually takes the period of time in which half of all the radioactive atoms disintegrate. This period is called the half-life and is a characteristic property of the given radioactive nuclide. The graph of Fig. 5 represents the rate of decay of a radioactive substance; the vertical scale gives the number of atoms which have survived up to the time given on the horizontal scale. This graph refers to the radioactive nuclide ^{210}Bi (radium-E), which has a half-life of 5 days. Starting off with 10,000 atoms, we find that after 5 days 5,000 atoms are left; after another 5 days there are 2,500 survivors; after another 5 days 1,250, and so on. This type of variation is called an exponential function and is characteristic for any process which occurs at random; in the case of radioactive disintegration this means that the decay of any one atom is independent of the fate of any other atom.

The radioactive substances occurring in nature have half-lives covering an enormous range, from thousands of millions of years to a fraction of a millionth of a second. The shortest half-life, 3×10^{-7} second, is of thorium-C′, the nuclide which emits the fastest a-particles; the longest half-life, $1 \cdot 4 \times 10^{10}$ years, is that of thorium-232, which emits the least energetic a-particles.

The radioactive strength of a given substance depends not on the total number of atoms in the substance but on the number of atoms which break up in a given time. A small amount of a substance of a short half-life may show a stronger radioactivity than a large amount of a long-lived substance, because in the former a larger proportion of the atoms break up every second; the effect of the latter will of course last longer. The number of atoms of a given substance which disintegrate every second is called the activity and is measured in a unit called the curie. It is nearly equal to the activity of one gramme of radium, and is the quantity of a radioactive substance in which 37 000 million atoms break up every second. The curie turned out to be a very large unit and for this reason smaller units, the millicurie (one thousandth of a curie) the microcurie (one millionth of a curie) the nanocurie (one thousandth of a millionth of a curie) and the picocurie (one millionth of a millionth of a curie) are often used. On the other hand, the huge quantities of radioactive substances produced in nuclear reactors, or released at the explosion of nuclear weapons, call for larger units of radioactivity, the kilo-curie (a thousand curies) and the megacurie (a million curies).

The Radioactive Series

As already stated, all radioactive nuclides among the heavy elements can be arranged in three series, which are shown in Fig. 6. The uranium series starts off with the nuclide uranium-238, which used to be called uranium-I and which has a very long half-life of $4\cdot5 \times 10^9$ years. It emits a-particles and is then transformed into an isotope of thorium of mass number 234, called uranium X_1. This has a half-life of only 24 days; it emits β-rays, and is transformed into an isotope of protactinium-234 called uranium-X_2, which has an even shorter half-life of $1\cdot2$ minutes. A second β-decay follows resulting in the production of another isotope of uranium of mass 234, called uranium-II. This sequence illustrates how a large number of nuclides can occur among a few elements: after one a- and two β-transformations we return to the same element. After uranium-II there is a sequence of two a-disintegrations leading to the formation of radium (^{226}Ra), which has a half-life of 1 600 years. The sequence from radium to radium-C has already been discussed. An inter-esting phenomenon occurs with this nuclide. It can decay either by a- or β-emission. In the first case, the product ^{210}Tl (radium-C'') emits a β-ray turning into an isotope of lead, ^{210}Pb (radium-D). In the second case, the product ^{214}Po (radium-C') emits an a-particle and is transformed into the same isotope, radium-D. Two other examples of such branching occur in radium-A and radium-E. Finally, the series ends with the formation of an isotope of lead of mass number 206. This nuclide was found to undergo no further transmutation; it is a non-radioactive, stable nuclide.

The second series, which is called the actinium series, starts off with an isotope of uranium of mass number 235, which used to be called actinouranium. It has a half-life 7×10^8 years and passes through a somewhat similar series of transformations, ending again in a stable isotope of lead, of mass number 207.

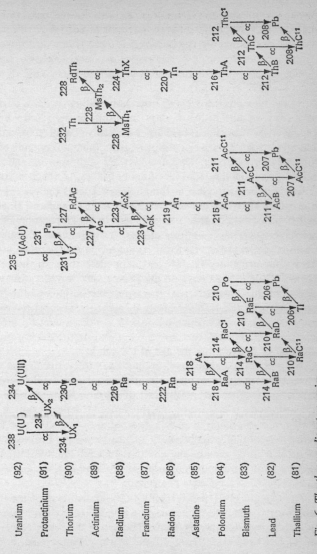

Fig. 6. The three radioactive series

Finally, the thorium series starts off with an isotope of thorium of mass 232, which has the very long half-life of 1.4×10^{10} years. As in the case of the uranium series there follow a number of transformations through successive a- and β-decays, ending finally in the formation of another stable isotope of lead, of mass number 208.

It will now be clear why we observe in nature nuclides of short half-lives, of the order of days, minutes or even fractions of seconds. These nuclides are being continually produced from their parents, which in turn are produced from the main parents of the three series, uranium-238, uranium-235 and thorium-232. All these three nuclides have very long half-lives, of the same order of magnitude as the age of the universe. This is the reason why they have survived. As will be shown later (Chapter 12), there exists a hypothesis that at one stage in the formation of the universe a very large number of radioactive nuclides, far greater than exist now, were created in a nuclear cataclysm from which all matter originated. If this had happened, then all the short-lived nuclides would have decayed since and only the ones with very long half-lives survived; these are the parents of the three radioactive series. It follows from this that the proportions of the radioactive elements which occur now are dependent on the age of the earth. In fact, this forms the basis for the most accurate method of determining the age of the earth.

If a mineral contains, say, uranium-238, then during the time that has elapsed since the formation of the mineral a certain amount of the stable end-product of uranium-238, i.e. lead-206, must have been produced, and this amount depends on the age of the mineral. Similarly, a certain amount of lead-207 must have been produced from the decay of uranium-235. From a comparison of the ratios of the amounts of lead-206 and 207, with the ratios of the amounts of uranium-238 and 235, the age of the mineral can be accurately determined.

Apart from the 41 nuclides forming the members of the three radioactive families, there are some isolated radioactive nuclides occurring among other elements. These are ^{40}K, ^{50}V, ^{87}Rb, ^{115}In, ^{123}Te, ^{138}La, ^{142}Ce, ^{144}Nd, ^{147}Sm, ^{148}Sm, ^{149}Sm, ^{152}Gd, ^{156}Dy, ^{176}Lu, ^{174}Hf, ^{187}Re, ^{190}Pt, ^{204}Pb. All of these are characterized by very long half-lives of the order of the age of the earth or much greater. The most interesting of these is potassium-40, which is the lightest of all very long-lived natural radioactive elements and a radioactive constituent of the human body. The total activity of radio-potassium in the body amounts to only 0·1 microcuries. A similar activity in the body arises from carbon-14, which has a much shorter half-life (about 6 000 years) but is being continually formed by the interaction of cosmic rays and the atmosphere.

Stable and Unstable Nuclei

So far we have described the main features of radioactivity as elucidated from a great many observations, but we have not yet attempted to find the reason for this phenomenon. Why are some elements unstable? Why do some substances emit α-particles, while others emit β-rays? In order to find an answer to these questions we must turn for a moment to stable nuclei.

We have already explained how, by means of the mass spectrometer, it is possible to analyse the isotopic composition of a given element. Such studies have shown that some elements, those with odd atomic numbers, have a very small number of isotopes, either one or two, while even elements have on the whole larger numbers of isotopes, up to 10. Altogether among the 81 elements with stable nuclides occurring in nature there is a total of 265 stable nuclides. Fig. 7 shows a chart of all nuclides occurring in nature. The horizontal scale gives the number of protons, and the vertical scale the number of neutrons. Each square represents one nuclide; the squares lying in one vertical line represent isotopes of the same chemical element. The stable nuclides are indicated by open squares; the radioactive nuclides by full squares.

It is seen that in light elements the number of protons is approximately equal to the number of neutrons, but as we go over to heavier elements the number of neutrons increases faster than the number of protons. In very heavy elements the ratio of the number of neutrons to protons is about 1·6. It is easy to explain the reason for this. Apart from the attractive nuclear forces acting between the nucleons there are also the electrostatic repulsive forces between the protons. Since the nuclear forces have a very short range they can only act between nucleons which are in the immediate vicinity of each other. The total attractive force will, therefore, be proportional to the number of nucleons in the nucleus. On the other hand, the electric force of repulsion between the protons has a long range, which means that each proton repels every other proton in the nucleus; the total repulsive force is thus approximately proportional to the *square* of the number of protons. As the number of protons in a nucleus increases, the repulsive forces increase faster than the attractive forces, and in order to restore the balance more nucleons are needed to provide attractive forces. Hence there must be more neutrons than protons.

It follows from this, as well as from other considerations, that only nuclei in which the ratio of protons to neutrons lies within certain limits can be stable. In fact it will be noticed that in the chart of Fig. 7 all stable nuclides occupy a very narrow region of the graph. What would happen if a nucleus were formed which lay above or below this region of stability? Such a nucleus would be unstable and would tend to undergo a transformation leading it into a more stable form. In most cases such transformation would be the conversion of a proton into a neutron or vice versa. We have already learned that such conversion can take place and that an electron and a neutrino are created in the process. Since electrons cannot exist in nuclei they are immediately ejected. We observe then the phenomenon of β-decay.

We can understand now why some nuclides are radioactive and undergo a β-decay, but we have still to explain a-decay; why do some elements which are apparently stable against β-decay break up with the emission of an a-particle? This appears to be particularly strange as the a-particle is not an elementary particle but an aggregate of 4 nucleons. In order to explain this it is necessary to consider the energy balance within the nucleus.

Binding Energy

When a nucleus is transformed into another, either by a- or β-decay, some energy is released in the process, which is taken up as the energy of motion of the a- or the β-particle (plus neutrino), with any surplus appearing as γ-rays. As we have seen, the

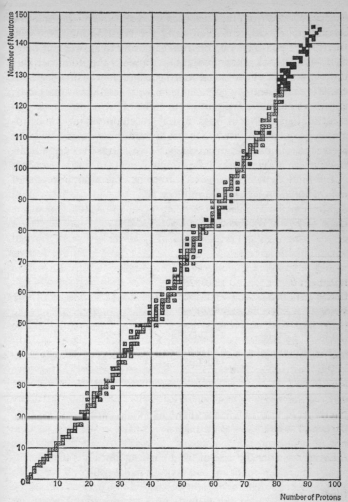

Fig. 7. Chart of nuclides occurring in nature

energy released from the nucleus may be quite high – of the order of millions of electron-volts – and the question may be asked where does this energy come from? In order to answer this question we must look more closely at the values of the atomic weights of nuclides. As already mentioned the atomic weights of all nuclides are very nearly whole numbers. Very nearly but not exactly, and this small difference turns out to be of great importance, so great that special techniques, mainly based on the mass spectrometer, have been devised to measure the atomic weights with the highest possible degree of accuracy. Table 2 gives the exact values, as far as we know them, of the atomic weights of several isotopes of the light elements, using as a reference the atomic weight of the carbon isotope of mass number 12.

Table 2. Atomic Weights of some Light Nuclides

Nuclide	Symbol	Atomic Weight
Neutron	^{1}n	1·0086654
Hydrogen	^{1}H	1·0078254
Deuterium	^{2}D	2·0141026
Tritium	^{3}T	3·0160501
Helium-3	^{3}He	3·0160301
Helium-4	^{4}He	4·0026033
Lithium-6	^{6}Li	6·0151238
Lithium-7	^{7}Li	7·0160053
Beryllium-9	^{9}Be	9·0121832
Carbon-12	^{12}C	12·0000000
Nitrogen-14	^{14}N	14·0030745
Oxygen-16	^{16}O	15·9949149
Oxygen-17	^{17}O	16·9991332

Let us consider the case of the α-particle. Since it is the nucleus of helium, and is made up of 2 protons and 2 neutrons, we might have expected that the weight of the helium nucleus would be equal to the sum of the weights of 2 protons and 2 neutrons. Actually this is not so. If we add up the atomic weights of 2 hydrogen atoms and 2 neutrons we obtain 4·0329816, while the atomic weight of helium is 4·0026033. We find thus that the weight of the helium nucleus is smaller than the atomic weight of its constituents by 0·0303783 units of atomic weight. (It is quite legitimate to use atomic weights when dealing with nuclei, because the extra weight of the electrons in the atom must balance on both sides of the equation.) What has happened to the missing weight? The answer is that it has been converted into energy in accordance with Einstein's principle of the equivalence of mass and energy. It is well known that in certain chemical reactions energy is released in the form of heat, and the amount of heat released is a measure of the stability of the given compound; because if we wanted to break up the compound we would have to deliver to it the same amount of energy that was released during its formation. Similarly, when a nucleus is formed from its constituents some energy is released, and this energy release is a measure of the stability of the given nucleus; for in order to break up the nucleus one would have to supply it with at least the same amount of energy. This is the reason why the compound nucleus weighs less than its constituents, and the difference, or mass defect, represents the binding energy of the nucleus. In the case of the helium nucleus the binding energy is 0·0303783 units of atomic weight. It is easy to convert this into energy units, say electron-volts. From Einstein's mass-energy relation (p. 135) it follows that one unit of atomic weight is equivalent to 931·48 MeV (million electron-volts). The binding energy of the helium nucleus is thus 28·297 MeV. This is the amount of energy which is released when an a-particle is formed, an amount which would have been obtained if a proton had been accelerated through a difference of potential of over 28 million volts. This gives an idea of the order of magnitude of the forces acting in the nucleus. This very high binding energy shows that the helium nucleus is a very stable structure.

In a similar fashion we can work out the binding energies for all other nuclides, and if these are plotted against the mass number we obtain a very interesting graph (Fig. 8). Actually, the graph of Fig. 8 shows not the total binding energy of the nucleus but the binding energy divided by the number of nucleons; it gives, therefore, the average amount of energy necessary to remove one particle from the nucleus. As is seen, the binding energy per nucleon starts off with low values in light elements

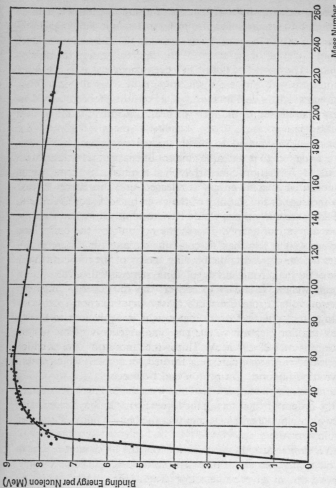

Fig. 8. Binding energy per nucleon

increases very rapidly, then flattens out and decreases again for heavier elements. In the region of light elements the variation is not as smooth as the continuous curve would lead one to believe. In fact some nuclides, like ^{4}He, ^{12}C or ^{16}O, lie well above the curve, which means that they are particularly stable structures.

There are two interesting features of this curve. First, for a large number of nuclides, with mass numbers from about 30 to about 120, the curve changes very little and gives a practically constant binding energy per nucleon of about 8·5 MeV. This means that for most nuclei the binding energy is proportional to the number of particles in it. This result is a strong argument in favour of the short-range character of the nuclear forces; for if they were of long-range, like electrical or gravitational forces, each nucleon would have interacted with every other and the total binding energy might have been expected to be nearly proportional to the square of the mass number. Secondly, the binding energy curve has a maximum in the region of medium-weight elements. This means that these elements of medium weight are the most stable ones. If we started with a very heavy element and split it into two, then some energy would be released, corresponding to the difference in the binding energy between medium and heavy elements. Similarly, if two light elements were fused together to form a medium weight element, there would again be a release of energy. Both these methods have actually been used in order to obtain nuclear energy for practical purposes, as will be discussed in the following chapters.

At the moment we are concerned with the problem of a-decay which occurs in the very heavy elements. In this region the average binding energy per nucleon is about 7·5 MeV; but from the slope of the graph (Fig. 8) we find that the binding energy per *additional* particle is only about 5·5 MeV. This means that if we wanted to take away one proton or one neutron from a heavy nucleus we would have to supply the nucleus with 5·5 MeV. Therefore, if we wanted to take away two protons and two neutrons separately we would have to supply the nucleus with a total energy of about 22 million electron-volts. On the other hand, we know that the binding energy of an a-particle is about 28 million electron-volts. Consequently, if these four particles came out not individually but combined as an α-particle, we would have a net gain of 6 million electron-volts, because we would have put in 22 MeV and received 28 MeV. Thus, although such a nucleus may be stable with regard to the emission of a proton or a neutron, yet it will be unstable with regard to the emission of an α-particle, because each time an α-particle is emitted we shall have a positive release of energy of about 6 MeV. This is the reason why heavy elements are unstable and break up with the emission of α-particles, which have an energy between 4 and 9 MeV.

Theory of α-decay

We have just shown that heavy nuclei are unstable because energy is released if they break up with the emission of α-particles. If so, the question may be asked why do they not break up instantaneously? Why does it take some nuclei many years to break up? In order to answer this we must discuss the variation of the potential energy inside and around the nucleus.

The positively charged protons inside the nucleus exert a repulsive force on any approaching particle with a positive charge; it is known from elementary physics that this force of repulsion increases as the distance between the charges decreases. A repulsive force corresponds to a positive potential energy, and the stronger the force the greater the potential energy. If, therefore, the potential energy between two positively charged particles is plotted as a function of the distance between them, a curve is obtained of the form BC (or B'C') in Fig. 9. In this graph the horizontal scale gives the distance of a positively charged particle from the centre of the nucleus, and the vertical scale the potential energy of the system. It is seen that as the particle approaches the nucleus from a distance, the potential energy is increasing, indicating a larger repulsive force. But at the point B (or B') the particle reaches the boundary of the nucleus and suddenly comes under the action of the very strong attractive nuclear forces. These attractive forces are much stronger than the repulsive electrical force, and consequently the potential energy suddenly drops to a large negative value.

The part of the curve BAA′B′ represents the shape of the potential energy inside the nucleus. Since we still do not know the nature of the nuclear forces, we cannot calculate the exact shape of the curve in the area BB′, but this does not really matter

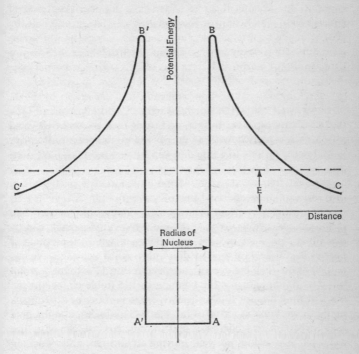

Fig. 9. Potential field around a nucleus as seen by a charged particle

for our present argument; for simplicity, it is shown as a rectangular well. The important point is that there exists a so-called potential barrier around the nucleus due to the action of the electric repulsive forces. Any positively charged particle coming from outside has to go over the top of the potential barrier in order to enter the nucleus, i.e. it must have an energy at least equal to the height of the barrier at B. Similarly, a particle inside the nucleus must cross the barrier in order to escape from it. It should be pointed out that the term barrier is used only in a figurative sense. There is no real barrier round the nucleus, but the electrical forces round the nucleus are such that if they were replaced by mechanical forces they would have been represented by a barrier. The height of the potential barrier depends on the charge of the nucleus and the approaching particle as well as on its radius; for a heavy nucleus and an a-particle it is of the order of 25 million electron-volts. As was shown before, an a-particle, if it could escape from the nucleus, would have an excess energy of about 6 MeV; this value, indicated on the graph by E, is, however, well below the potential barrier, and consequently the particle cannot escape. In fact, according to the laws of classical physics, an a-particle with an energy below the potential barrier would never be able to come out, and so we would not have been able to observe a-radioactivity.

The fact that we do observe the emission of a-particles can only be explained on the basis of wave-mechanics, which tells us that radiation often behaves like matter and matter like radiation. According to this, the motion of an a-particle can be described as a wave motion, and the space within the potential barrier as a nearly opaque medium, through which the wave has to penetrate. The chance of this penetration is extremely small but it is finite, and like the winning of the treble chance in the pools it may happen if a sufficiently large number of attempts is made. If we think in terms of mechanical analogies, and imagine that there is a real barrier round the nucleus, we can visualize the a-particle inside the well, moving to and fro, and knocking on the wall of the barrier. In the vast majority of these collisions the a-particle is reflected back, but there is a very tiny probability, of the order of one in 10^{24}, of it penetrating through the barrier and emerging outside. The escape of a-particles is thus a matter of chance, and this is the reason why in a given substance some atoms decay very rapidly, while others exist for a very long time. The probability of escaping through the barrier will, of course, depend on the thickness of the wall to be penetrated. An a-particle which has a large energy will have to go through a thinner barrier and, therefore, its chances of escape are much greater. On the basis of this reasoning we may expect that radioactive substances which send out a-particles of high energy should have a shorter half-life, because the a-particle will be able to get through quickly, while those of a low energy will have a longer half-life. In fact, such a relationship between the energy of the a-particle and the half-life of the substance has been known for a long time and called the 'Geiger-Nuttall Rule'. It was an empirical relationship which was not understood until the theory of a-decay, just outlined, was put forward in 1928 by Gamov. This theory gave not only a qualitative but also a quantitative account of all observed facts. We can, therefore, say that we understand now why the heavy elements decay with the emission of a-particles and why some of them have long and others short half-lives. We also understand why the Periodic Table of naturally occurring elements finishes at the ninety-second element; elements heavier than uranium are expected to have half-lives short compared with the age of the earth.

Theory of β-decay

In the case of β-decay, the theory is somewhat different as we deal here not with the emission of nuclear constituents but with the transformation of one nucleon into another, and, resulting from it, the emission of an electron and neutrino. By considering the interaction between the nucleons and the electron plus neutrino which are formed during this process, it is possible to calculate the probability of a given nucleus undergoing β-decay. As was shown in Chapter 1, this theory gave a good explanation of the experimental facts.

It should now also be clear why β-emission is observed after a-emissions among the heavy elements. After several successive a-transformations the residual nucleus differs from the original one by the loss of an equal number of neutrons and protons. Since in the region of heavy elements the stable nuclei contain many more neutrons than protons, the loss of the same number of neutrons and protons means that in the new nucleus the ratio of neutrons to protons is higher than that corresponding to a stable nucleus. This can be adjusted by one of the neutrons being transformed into a proton, with the simultaneous emission of an electron and a neutrino.

Artificial Radioactivity

The chart of nuclides shown on Fig. 7 represents all nuclei which occur in nature in detectable quantities. Among these are the naturally occurring radioactive nuclides, shown by black squares. The remainder are the 265 stable nuclides. Since a very careful search has failed to reveal the presence of any other type of nuclei in nature we may conclude that these are the only stable nuclei. Any other nucleus of a different constitution from those is, therefore, probably unstable.

By means of nuclear disintegrations, which will be discussed in the next chapter, it is now possible to produce nuclei other than those shown in the chart and it has been found that all these are indeed radioactive. In general, those of the artificially produced nuclei which have a higher than normal neutron to proton ratio decay by emission of electrons; those in which this ratio is lower decay by emission of positrons. In many cases an alternative to positron emission is the so-called K-capture, when the nucleus captures one of its orbital electrons, usually the nearest, which is called a K-electron. In a few cases artificially produced a-emitting nuclides have been observed, particularly among the heavy elements. In fact, a fourth radioactive series, which does not exist in nature, has been observed, starting with the artificially produced element neptunium-237 (Fig. 10).

Altogether by various means, which will be described later, over 1 200 radioactive nuclides have now been identified, nearly five times the number of stable nuclides. All these are shown on the chart of Fig. 11, which is similar to that of Fig. 7; the open squares represent stable nuclides and the black squares radioactive ones. There is now at least one radioactive isotope known of any of the existing chemical elements, and radioactive isotopes have been produced of those elements which do not occur in nature, such as technetium (43) or promethium (61). In fact, a number of isotopes have been produced of elements heavier than uranium, namely neptunium (93), plutonium (94), americium (95), curium (96), berkelium (97), californium (98), einsteinium (99), fermium (100), mendelevium (101), nobelium (102), lawrencium (103), rutherfordium (104) and hahnium (105). A search is now being made, using heavy-ion accelerators, for still heavier elements, in particular, for elements around atomic

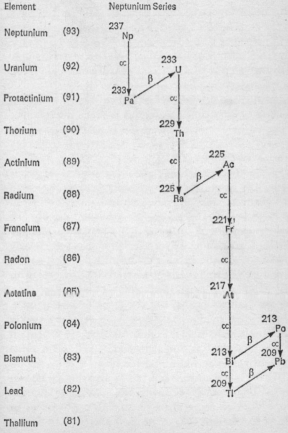

Fig. 10. The neptunium series

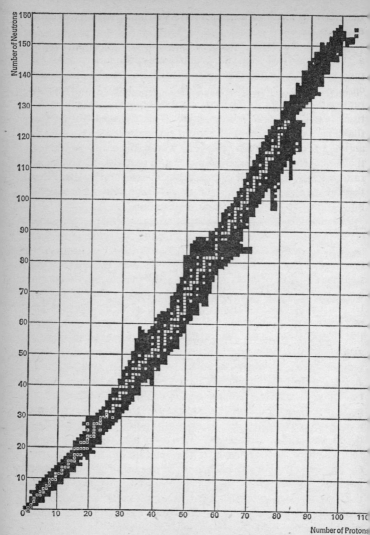

Fig. 11. Chart of all known nuclides

number 114, which are expected to be less unstable and therefore easier to discover.

The fact that every chemical element can be produced in a radioactive form has important implications not only for physics but for many other fields of science as well as for medicine, technology and industry. One of the many ways of using radioactive isotopes is the tracer method in which minute quantities of radioisotopes are employed. This is based on the fact that a radioactive isotope of a given element behaves chemically and biologically in exactly the same way as the stable isotope of the same element. The only difference is that the possession of radioactivity labels it, making possible its detection by observing the radiations emitted. This can be achieved with an extremely high sensitivity by using a counting technique.

The simplest of these is the Geiger counter, which in principle consists of a tube with a wire along its axis, and filled with a gas or a mixture of gases under specified pressure. A strong electric field is set up between the walls of the tube and the central wire, such that if an α- or β-particle entered the tube, it would set up a momentary discharge resulting in a voltage pulse which can be registered by means of any counting device, say a telephone meter. In this way the number of electrons which pass through the Geiger counter in a given time can be recorded. An activity as low as a few disintegrations per second, i.e. about a hundred picocuries, can be easily detected. For a medium weight radioactive nuclide of a half-life of about 10 days, this activity would be obtained from an amount of material weighing about 10^{-15} grammes. Thus by means of radioactivity it is possible to detect a quantity of a given material millions of times smaller than that detectable by any other method. Even more sensitive, particularly for the detection of γ-rays, is the scintillation counter. This contains a phosphor, which may be a crystal of sodium iodide activated with thallium. The ionizing radiation passing through the crystal produces flashes of light which enter into a photomultiplier; there they are converted into electron pulses which can be registered.

The usual procedure in the tracer method is to add a small amount of a radioactive isotope to the inert substance whose behaviour or history in any technical or biological process it is desired to follow. In industry, for example, this method has been applied to measure the amount of wear and tear of piston rings in engines, to detect leaks in underground water pipes, or to check the flow of petrol in pipelines. In agriculture isotopes have been used to study the utilization of fertilizers, the spread of plant diseases, or the uptake of various materials by plants. Finally, in medicine radioisotopes have found very wide applications, both for the fundamental study of the processes going on in the living organism and as an aid in the diagnosis of disease.

Another field of application utilizes the effects produced by the radiations themselves. In this case use is made of the fact that quantities of radioactivity exceeding thousands of times those from the available natural radioactive elements can now be produced. In some cases radioactive isotopes having activities of

Table 3. Some Radioactive Nuclides used in Medicine

Nuclide	Symbol	Half-life	Type of radiation emitted
Tritium	^3T	12·33 y	e^-
Carbon-14	^{14}C	5 692 y	e^-
Sodium-24	^{24}Na	15·0 h	e^-, γ
Phosphorus-32	^{32}P	14·3 d	e^-
Sulphur-35	^{35}S	87·2 d	e^-
Chromium-51	^{51}Cr	27·7 d	γ
Iron-59	^{59}Fe	45 d	e^-, γ
Cobalt-57	^{57}Co	271 d	γ
Cobalt-58	^{58}Co	71·4 d	e^+, γ
Cobalt-60	^{60}Co	5·27 y	e^-, γ
Selenium-75	^{75}Se	120 d	γ
Krypton-85	^{85}Kr	10·73 y	e^-, γ
Strontium-85	^{85}Sr	65·2 d	γ
Technetium-99m	$^{99}Tc^m$	6·0 h	γ
Iodine-125	^{125}I	59·9 d	γ
Iodine-131	^{131}I	8·06 d	e^-, γ
Gold-198	^{198}Au	64·7 h	e^-, γ
Mercury-197	^{197}Hg	64·1 h	γ

kilocuries can be easily manufactured. In fact, enormous quantities of radioactive substances are created as a by-product of atomic energy. The problem of disposal of these materials without polluting the environment and causing harm to living organisms may turn out to be very difficult to solve.

Large quantities of radioactive nuclides – but still only a tiny proportion of the amount produced – can be utilized in various applications, such as sterilization of pharmaceutical preparations, and, in some cases, of food products; as sources of power in satellites and space vehicles, or of light; as sources of electricity in heart pacemakers; and, above all, as sources of radiation for the treatment of cancer.

Not all radioisotopes can be used in these various applications. The half-life, the type of radiations emitted, the specific activity and availability have to be taken into account in the choice of radioactive isotopes. Table 3 gives a list of the radioisotopes most frequently employed in medicine.

It is said that the possibilities opened to research in technology and medicine by the availability of these artificially produced radioactive isotopes may equal in importance the use of nuclear energy as a source of power. It has certainly already helped to make remarkable progress in a number of branches of science and technology.

CHAPTER 3

Smashing the Atom

The Nucleus as a Target

The discovery of radioactivity revealed for the first time the existence of enormous stores of energy within the atom. The energies of the particles emitted from radioactive substances are of the order of millions of electron-volts, as compared with a few electron-volts which is the energy released in atomic processes, such as chemical reactions. This discovery immediately raised the question whether it would be possible to make practical use of this huge store of energy. Apart, however, from this utilitarian motive, there is the natural urge to find out all about the world around us, in this case, about the nucleus of the atom. Although much can be learned about the nucleus and its properties by the passive observation of the processes which nature displays for us in the phenomenon of radioactivity, it is clear that in order to gain a better insight into this subject a more active approach is needed. Furthermore, natural radioactivity is practically limited to heavy elements; of much greater value would be to know about the structure of light nuclei, since these are simpler systems lending themselves more easily to theoretical analysis.

The question then is how to get at the nucleus. Apart from being extremely small it is also very inaccessible; it has a potential barrier around it which prevents the approach of charged particles. The only way of getting to know about the nucleus seems to be by storm, by trying to break it up. The nucleus has often been likened to a little inaccessible island, surrounded by a high wall, and with heavy cloud above it. The only possible way to find out something about the island and its inhabitants is to throw stones into it over the wall. By observing the directions in which the stones are hurled back one might make a guess about the intelligence of the inhabitants; if sometimes a different type of stone is thrown back it is possible to infer something about the mineral composition of the island. Similarly, in the case of the nucleus the only possible way to find out about its structure is to bombard it with some suitable projectile. In the days when Rutherford began his experiments the only suitable projectiles which had sufficient energy to penetrate the potential barrier were the a-particles emitted from radioactive substances. These were, therefore, the first to be used for the artificial disintegration of nuclei.

Before we discuss the results of such disintegrations we ought to make an attempt to calculate the chance of hitting a nucleus and producing a disintegration. First of all, we cannot aim at the target and we have to shoot at random. Secondly, the target is extremely small, the diameter of the nucleus being about ten thousand times smaller than the diameter of the atom. Thus, if a beam of a-particles is made to pass through matter, the chance of any one of them scoring a direct hit on the nucleus is very small indeed. If we consider the nucleus as a solid target the probability of hitting it would be proportional to the cross-section of the nucleus. Actually this probability depends on a number of other factors, such as the potential barrier around the nucleus, and the wave properties of the particles which may give rise to certain 'interference' effects and make the probability of hitting the nucleus either much smaller or much larger than the geometrical cross-section. Nevertheless, we still express this probability in terms of a cross-section. The unit for this cross-section is called a barn, which has the jocular origin 'as big as a barn'. One barn is an area of 10^{-24} cm^2, which would be roughly the cross-sectional area of a particle of radius about 6×10^{-13} cm. The relative areas of the nucleus and the atom are roughly the same as those of the *Queen Elizabeth* and the North Sea.

Small as the nuclear target may be, there are so many atoms in even a minute amount of material that if our projectile were allowed to proceed long enough it would be bound in the end to hit a nucleus. Unfortunately, the projectile, when it is a charged particle like the a-particle, is not allowed to proceed a long distance unmolested. As it penetrates through matter it interacts with a very large number of electrons of the atoms passed by. Since electrons have a very small mass as compared with that of the a-particle, they may be thrown out from their atoms even when the a-particle passes at a relatively large distance from them. This process is called ionization. The amount of energy lost by an a-particle at each such ionization is very small – in air it is only about 34 electron-volts – but these ionizations are so numerous that the energy of the a-particle is very quickly used up. Thus, in passing through air at atmospheric pressure an a-particle may make about thirty thousand ionizing collisions along each centimetre of its path, losing in this way an energy of over 1 MeV. The energy of the a-particle is, therefore, very quickly dissipated and the particle is brought to rest before it has had a chance of making a collision with a nucleus. We see now why the range of an a-particle in air is only a few centimetres and why the track of an a-particle is straight (Plate VI); each of the ionizing events is far too feeble to deviate it from its course. It is only very rarely that an a-particle comes close enough to a nucleus to be deflected from its straight path by the repulsion between like charges, but occasionally this is observed. The track shown in the photograph of Plate VII shows such scattering events. This a-particle underwent two nuclear collisions: in one it was deflected only slightly from its course; in the other, near the end of its range, it suffered a much stronger deflection. In fact, it was the observation of the scattering of a-particles that led Rutherford to the concept of the atomic nucleus.

If, however, ordinary scattering is rare, the actual entry of an a-particle into the nucleus to cause its disintegration is a rarer event still. The reason for this is that even if an a-particle happens to score a direct hit on the nucleus it will not be able to enter into it unless it has a sufficient amount of energy. This is due to the existence of the potential barrier around the nucleus which was discussed in the previous chapter (Fig. 9). The graph of Fig. 9 shows the shape of the potential field as seen by an approaching charged particle. As the particle gets nearer to the nucleus the repulsive force acting between two positive charges increases. The only way of overcoming this repulsion is to use a particle of an energy higher than that corresponding to the top of the potential barrier. The height of the barrier depends on the charge of the target and of the projectile; Fig. 12 shows the

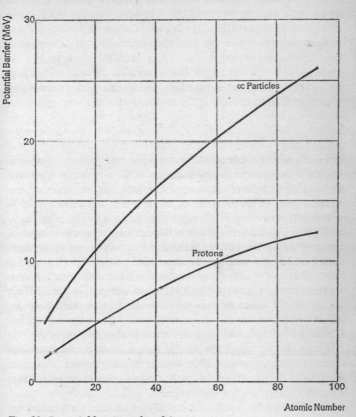

Fig. 12. Potential barriers of nuclei

variation of the height of the barrier with the atomic number of the element for various projectiles. For *a*-particles and heavy nuclei it is about 25 MeV; in light nuclei it is much lower, for nitrogen it is about 5 MeV. As explained in the previous chapter, the projectile need not necessarily cross over the top of the barrier in order to enter the nucleus; its wave properties enable it to penetrate through the barrier. But the probability of such penetration decreases very rapidly with energy, and a particle of an energy considerably lower than the barrier height will have an extremely small chance of getting into the nucleus.

Summing up, there are two chief requisites for a successful nuclear disintegration: a very large number of projectiles, to ensure that a few will be able to score a hit before having dissipated their energy on ionization, and a high enough energy of the projectiles to overcome the potential barrier.

The α-particle as a Projectile

On the basis of these considerations the odds in favour of producing a disintegration by means of the *a*-particles from the radioactive substances appear quite low, for we are then restricted both in the number of the particles, as well as in their energy. Nevertheless, Rutherford carried out such an experiment in 1919 and was successful; he produced the first artificial nuclear disintegration. The target element was nitrogen, which Rutherford bombarded with *a*-particles from radium-C′ (^{214}Po). He found that as a result of this bombardment protons were emitted. In this experiment the *a*-particle source was put in a tube from which the air could be removed. At one end of the tube there was a thin metal foil but sufficiently thick to stop the *a*-particles. On admitting nitrogen into the tube Rutherford observed the emission of particles through the foil, and these particles were identified as fast protons. The number of these protons was quite small, and it was calculated that only one *a*-particle in 100,000 succeeded in producing this reaction.

From the observation of the emission of protons Rutherford deduced that the following reaction had taken place:

$$^{14}N + {}^4He = {}^{17}O + {}^1H.$$

In other words, he postulated the transmutation of nitrogen and helium into oxygen and hydrogen.

This equation, written in the form of a chemical reaction,

represents the first transmutation of elements performed by man. This was an event of such great importance that it is worth while to analyse it in greater detail. The reader may ask how we know that it was this and not some other reaction which took place. Since neither the oxygen nor the hydrogen were formed in sufficient quantity to be collected, how could one be sure about the products of this reaction?

The first thing to check about a nuclear reaction is that the total charge and the total number of particles remain unchanged, i.e. the number of protons and the number of neutrons must be the same on both sides of the equation. The symbol ^{14}N stands for 7 protons and 7 neutrons; therefore, the sum of particles on the left is 9 protons and 9 neutrons. ^{17}O has 8 protons and 9 neutrons; therefore, together with the emitted proton we have again 9 protons and 9 neutrons. This check by itself is, however, not sufficient, because we could write down several other equations which fulfil the same conditions. How do we know which reaction has taken place? This can be answered by making use of two fundamental laws of classical physics, the laws of conservation of energy and of momentum. In any elastic collision, say between two billiard balls, the total energy and the total momentum before and after the collision must be the same. With two identical balls, if one ball makes a head-on collision with another at rest, the first will come to a stop and the second will carry on in the forward direction with the energy and momentum previously had by the first ball. In off-centre collisions both balls will be set into motion at right angles to each other, and the direction and energy of each of the balls can be simply calculated from the conservation laws. Similarly, if an a-particle collides with a helium nucleus both are set in motion at right angles to each other. We can often observe the paths of both particles after the collision in a cloud chamber and verify that the laws of conservation of energy and momentum hold for nuclear processes. Such a cloud-chamber photograph is shown on Plate VIII. In this case the a-particles were allowed to pass through a chamber filled with helium and a number of their tracks are seen. One of them is seen to fork producing two tracks at right angles to each other. By measuring the lengths of the three tracks in the fork we can determine the energies of the particles involved and their momenta. In this way it was possible to verify that the laws of conservation of energy and momentum hold strictly in nuclear collisions.

The same laws can be applied to the disintegration of nitrogen, but since this is not an elastic collision we have to consider the energy balance of the reaction. The target, the nitrogen nucleus, is initially at rest; the projectile, the a-particle, has a known energy. But this is not the total energy involved. As was shown in the previous chapter, each nucleus has a certain binding energy; the binding energies of the products of a nuclear reaction will on the whole be different from the binding energies of the initial nuclei. This means that in each nuclear reaction a certain amount of energy may be either released or absorbed, depending on whether the products are more or less stable than the original nuclei. This is taken into account by adding to the equation the so-called Q-value. The equation for the disintegration of nitrogen will, therefore, read:

$$^{14}N + {}^{4}He = {}^{17}O + {}^{1}H + Q.$$

The Q-value is a measure of the change in the binding energy resulting from the reaction. As has already been explained, the binding energy is obtained from the masses of the particles involved, and so the Q-value can be simply calculated from a knowledge of the atomic weights of all the nuclides involved. It is equal to the difference in the masses on both sides of the equation. Let us carry out the calculation (see Table 2). The atomic weights of nitrogen-14 and helium-4 are 14·0030745 and 4·0026033, giving a sum of 18·0056778. The atomic weights of oxygen-17 and of hydrogen are 16·9991332 and 1·0078254, which add up to 18·0069586. We see that the sum on the left is smaller than the sum on the right, that is to say, we have produced particles heavier than the ones with which we started by 0·0012808 units of atomic weight, which corresponds to an energy of 1·193 MeV. This, therefore, is the Q-value of the reaction. It is negative in this case, which means that this amount of energy has been lost in bringing the reaction about. This, of course, was to be expected when we remember that the a-particle is a very stable structure, and when it is broken up to take a proton from it, energy must be supplied. This energy has to come from the kinetic energy of motion of the a-particle which produces the reaction.

If this is taken into account, we know now how much energy is left over to be shared between the products ^{17}O and the proton. By applying the laws of conservation of momentum we can then calculate the proportion in which this energy is shared between the two products for any direction in which they may be emitted. Thus, in order to identify the reaction we have to measure the energies of the particles and the angles at which they are emitted in relation to the original direction of the a-particle and check whether they agree with the calculated values. This can again be done by means of the cloud chamber; the photograph of Plate IX shows an actual process of disintegration of nitrogen. We see there tracks of a-particles passing through a chamber filled with nitrogen. At one point a fork occurs in one of the tracks; the short track to the right is that of the oxygen nucleus formed, and the thin long track to the left of the emitted proton. From the ranges of these particles their energies can be determined; the angles are read off directly. It is found that the values thus obtained fit excellently with the calculated ones. This is a very sensitive test and can only be fulfilled for one type of reaction. Thus, although we are unable to see the particles themselves, such a picture gives us as good evidence of this reaction having occurred as any seen directly.

Soon after Rutherford's experiment with nitrogen similar reactions were produced in many other elements. Some were of the same type, i.e. leading to the emission of protons, but other types of reaction were also observed; two of these, which have historical interest as being the first of their class, will be discussed now.

One of them is the emission of neutrons as a result of the bombardment of beryllium with a-particles. The formula of the reaction is

$$^{9}Be + {}^{4}He = {}^{12}C + {}^{1}n.$$

It was this reaction which led to the discovery of the neutron in 1932 by Chadwick, as described in Chapter 1.

The Q-value of this reaction is $+5.702$ MeV, as can be verified by calculating the atomic weights of the particles involved given in Table 2. In this case there is, therefore, a positive release of energy: most of it is taken up by the neutron which is, therefore, emitted with a large amount of energy. Although this reaction cannot be observed directly in the cloud chamber, since the neutron as a non-ionizing particle does not produce a cloud chamber track, it is possible sometimes to observe the track of the carbon-12 nucleus, and from this to deduce the direction and energy of the neutron. Once again the values are in complete agreement with those calculated on the basis of the laws of conservation of energy and momentum.

Neutrons themselves are best detected by making them collide with protons, i.e. by letting the neutron beam pass through a hydrogenous medium, say water or paraffin wax. Since the neutron and the proton have approximately the same mass their collision is analogous to that between two billiard balls. In a head-on collision the neutron is brought to rest and the proton is emitted in the forward direction with the whole energy. In off-centre collisions the proton may be emitted at different angles but from the angle of emission and from the energy of the proton the energy of the neutron can be deduced.

The second type of reaction is that which led for the first time to the production of artificial radioactive isotopes. It was discovered by Irene Curie and Frederic Joliot who carried out experiments similar to Rutherford's but using aluminium as the target. They used a Geiger counter to detect the particles emitted, and noticed that the aluminium continued giving off radiations even after the bombarding projectiles, the α-particles, were removed. They soon established that this radiation was of the β-type and that it was decaying gradually with a half-life of about three minutes. They interpreted this phenomenon as follows: as a result of the bombardment of aluminium with α-particles, an isotope of phosphorus of mass number 30 and a neutron are produced according to the equation:

$$^{27}\text{Al} + {}^{4}\text{He} = {}^{30}\text{P} + {}^{1}\text{n}.$$

The product nucleus, phosphorus-30, is not a nuclide found in nature. It contains 15 protons and 15 neutrons, but for so large a nucleus more neutrons than protons are needed to make it stable; this nuclide is consequently unstable, and the transition to stability can be achieved by the transformation of one of its protons into a neutron, with the simultaneous emission of a positive electron. Thus, the above reaction is followed after a time by the reaction

$$^{30}P \rightarrow {}^{30}Si + e^+.$$

Immediately after this discovery was made in 1934 it was found that, far from being an exceptional event, such nuclear disintegrations in which unstable nuclides are produced occur very frequently.

Artificial Projectiles

All disintegrations described so far were produced by using a-particles as projectiles. It is clear, however, that these particles offer a rather limited scope for study. First, the number of projectiles is limited by the amount of radioactive substance available and this very seldom exceeded one gramme of radium. Taking into account that some nuclear reactions are produced at a rate of one for several million a-particles, it is obvious that one needs a very intense source of radiation to study the reactions in detail. Even more important, however, is the fact that the energy of the a-particles from the radioactive elements is limited; the maximum energy is less than 9 MeV. This means that it is practically impossible to produce disintegrations in heavy nuclei, where the potential barrier is much higher, reaching 25 MeV (Fig. 12). On the other hand, the barrier for singly charged particles, e.g. protons, is much lower, and an energy of 12 million electron-volts would be sufficient to break up even a heavy nucleus. It seems, therefore, that a good case is made out for using projectiles made in the laboratory, if sufficiently intense beams of such particles could be produced. What is needed is a device in which protons or other charged particles would be accelerated to energies of the order of 10–20 million electron-volts. This was achieved in the various accelerators already referred to, of which there is now a very imposing array.

It started off in a very modest way with an energy of only 600 000 electron-volts. Although this energy is well below the potential barrier, even for light elements, it was thought that if a sufficiently large number of such low-energy protons were available, a small fraction of them might be able to penetrate the barrier and produce the disintegration. This was the idea in the minds of Cockcroft and Walton who built in 1932 a fairly simple high-voltage installation, using a transformer, rectifiers and condensers, to produce a potential of about 600 000 volts. By producing a beam of hydrogen ions and letting it pass through this potential difference they accelerated protons to an energy of 600 000 eV.

These protons were then used to bombard a lithium target. The result was the emission from lithium of a-particles of very high energy, according to the reaction

$$^7\text{Li} + {}^1\text{H} = {}^4\text{He} + {}^4\text{He}.$$

If we put in the values of the atomic weights (Table 2) we find that the Q-value is in this case 17·35 MeV. This large amount of energy released in the reaction appears as the kinetic energy of the two a-particles, which are fired away in opposite directions. The tracks of these a-particles can easily be observed in a cloud chamber, and in this way the occurrence of this first nuclear disintegration produced entirely by artificial means could be confirmed.

This reaction is also important for the reason that it was the first to produce a large release of nuclear energy; we put in 0·6 MeV and obtain over 17 MeV, nearly a thirty-fold gain. All the same it has no practical value as a source of energy, for the overall balance is strongly negative. This is so because the yield of this reaction is extremely small. At these low bombarding energies perhaps only one out of one hundred million protons would be able to penetrate the barrier to produce this reaction; all other protons lose their energy on ionization.

The reason why we observe the reaction at all, despite its low yield, is that we start off with an enormous number of projectiles. The beam of protons produced in an accelerator constitutes an electric current which in modern accelerators may be as much as many amperes, and even a current of one microampere corresponds to the emission of 6.2×10^{12} protons per second, which is equivalent to about 170 grammes of radium. This illustrates the great advantage gained by using artificially produced projectiles. If this could be coupled with high energy as well, one might expect very impressive results. This is the reason why immediately after Cockcroft's experiments there was such a rush to build accelerators to produce intense beams of high-energy particles.

These particles are either protons which are obtained from hydrogen, or artificial a-particles which are obtained from helium. But the most effective of artificial projectiles turned out to be deuterons. The deuteron is the nucleus of deuterium (symbol D), an isotope of hydrogen of mass 2, and is composed of one proton and one neutron. The abundance of deuterium in nature is relatively low, about one out of 7 000 hydrogen atoms is that of deuterium; nevertheless it is possible to separate it and obtain it in practically pure form. It is usually obtained as heavy water (deuterium oxide), which differs from ordinary water by the substitution of ordinary hydrogen by deuterium.

Deuterons have proved to be more effective projectiles than protons for the following reason. A deuteron passing near a nucleus may split up in flight; its neutron is captured by the nucleus and the proton flies on. Since a neutron has no charge there is no potential barrier for it and consequently such reaction can occur even with a projectile of a fairly low energy. Many of the nuclear reactions leading to the production of radioisotopes have been produced by means of the bombardment of various elements with deuterons.

Another frequent use of deuterons is as a source of neutrons. One of the best sources is the bombardment of deuterium with deuterons, according to the reaction

$$^2D + {}^2D = {}^3He + {}^1n.$$

Particle Accelerators

One of the first accelerators to be developed was the electrostatic generator, or Van de Graaff machine, after the name of its inventor. It is a straightforward electrostatic machine in which a conductor is charged up to a high potential. Its principle is illustrated in Fig. 13. A spherical conductor is electrically insulated from the ground by means of a long cylinder of insulating material. A silk or rubber endless belt runs on pulleys from the bottom of the cylinder, which is at ground potential, to the top terminal. A positive electric charge is sprayed upon the belt

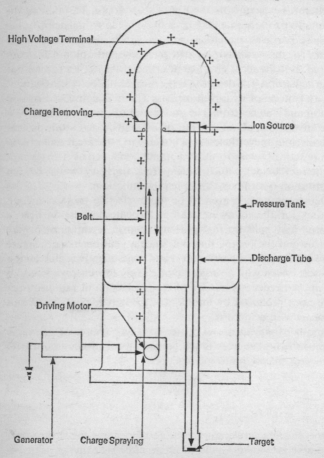

Fig. 13. Principle of the Van de Graaff generator

which carries it to the upper conductor. In this way the conductor gradually receives a larger and larger charge and its potential goes up correspondingly. In principle it should be possible to raise the potential to any desired value but in practice there is a limit set by the breaking down of the insulation properties of the air, resulting either in a spark discharge or in a corona discharge on the surface of the conductor. These factors limit the potential obtainable to about 2 million volts. A higher potential can be achieved if the whole machine is enclosed in a tank filled with a gas at high pressure, preferably a gas in which discharges do not occur easily, such as nitrogen or 'Freon' (CCl_2F_2). In this way, it is possible to increase considerably the final potential.

To produce the beam of projectiles a discharge tube is provided containing a hydrogen ion source at the top terminal. The protons or deuterons passing through the tube to the bottom are accelerated to the high energy. Several such generators, producing beams of particles up to about 4 MeV, are in existence in various countries. This seems to be about the limit obtainable in this type of generator. A doubling of the energy of the particles with the same accelerating potential can be achieved in the so-called 'tandem generator', which is a modification of the Van de Graaff machine. In the tandem generator negative ions, e.g., hydrogen atoms which have picked up an additional electron, are first accelerated from ground potential to that produced by the machine. By stripping off some electrons the negative ions are converted into positive ions and these are then accelerated back to ground potential. In this way it became possible to obtain energies greater than 25 million electron-volts. If instead of hydrogen a heavy element, such as iodine, is used, with many more electrons to strip, energies of several hundred million electron volts may be achieved.

Although the Van de Graaff machine has many advantages, for it provides a fairly high current at a very well-defined and controllable energy, yet its inherent limitations make it necessary to look for other accelerators.

One of the most successful accelerators in the field of nuclear disintegrations is the cyclotron invented by Lawrence in 1930. This is a machine in which particles are accelerated to high energies without actually using high potentials. The principle of the cyclotron is explained in Fig. 14. A copper box made up of two hollow semi-circular electrodes is situated between the poles of a powerful magnet. The two dees, as they are called after their shape, are connected to a source of potential, say about 100 000 volts, alternating at a high frequency. An ion source in the centre of the box provides charged particles, say protons. The protons, being positively charged, will be accelerated towards one of the dees which at that instant happens to be negatively charged. In crossing the gap they will thus gain an energy of

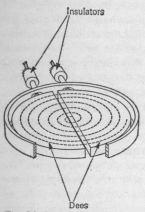

Insulators

Dees

Fig. 14. Principle of the cyclotron

100 000 electron-volts. Since all this happens in a magnetic field the protons moving into the dee will describe a semi-circular path and at a certain time afterwards will again reach the gap. By synchronizing the frequency of the voltage applied to the dees with the time it takes for the particle to describe a semi-circle it can be so arranged that the protons will reach the gap after the potential of the dees has changed signs. The protons will, therefore, receive another acceleration from the same voltage, and having now a higher energy, 200 000 eV, will move in a circle of a larger radius. Although their path is now longer their

velocity goes up in the same proportion and so it will take them the same time to reach the gap again; by that time the voltage will be reversed again and so the protons receive a further acceleration. This process is repeated many times; every time the protons pass the gap they receive an acceleration of 100 000 eV. The radius of their path gradually increases until they reach the edge of the box. By that time they may have been accelerated about a hundred times so that their total energy will be 10 MeV. Thus by using a potential of only 100 000 volts, one obtains particles of an energy corresponding to a hundred times greater potential.

Actually the final energy of the particles is independent of the voltage applied to the dees, and is determined entirely by the radius of the magnet and the intensity of the magnetic field. Consequently, by increasing the size of the poles and the magnetic field, very high energies can be obtained. For example, using a magnet with poles 1·5 metres in diameter, deuteron beams can be accelerated to an energy of 24 MeV.

In this type of cyclotron there is a limit to the energy, due to the increase of mass with velocity which follows from the theory of relativity (see Chapter 6). As the velocity of the particles increases so does their mass and consequently the time it takes to describe the circular path becomes longer, and the particles get out of phase. In fact, the example given in the previous paragraph represents the maximum energy that can be obtained with the cyclotron.

The limit imposed by the relativistic increase of mass has been overcome after the discovery of the principle of phase stability. This shows that the particles in a cyclotron tend to bunch in certain orbits on which they go round and round without gaining or losing energy as long as the frequency of the dee voltage remains constant. If, however, the frequency is decreased, the whole bunch experiences an accelerating field and moves to a new stable orbit. The lower frequency matches the longer time it takes for particles of a higher energy to go round. In this way, by gradually decreasing the frequency, the energy of the particles is gradually increased.

A cyclotron modified in this way is called a synchro-cyclotron, and it can be used to accelerate particles to very high energies. However, unlike the conventional cyclotron, in which a continuous beam of particles is produced, in the synchro-cyclotron the particles come out in bunches, which means that the average current is greatly reduced. All the same, the gain in energy far outweighs this loss. A number of synchro-cyclotrons are now in existence; the first of them, at Berkeley, California, had a pole diameter of 4·7 metres and produced beams of protons of an energy 730 MeV.

The reader may ask why we need to go to such high energies, of the order of 1 000 MeV or 1 GeV (1 Giga electron-volt = 10^9 eV), when about 30 MeV should be sufficient to climb over the highest potential barrier? There are several reasons for this. First, by using much higher energies we are able not just to chip off one or two particles from the nucleu but to break it into a large number of fragments. Such disintegration is called spallation and leads to many new type of unstable nuclei. Secondly, in order to study the nature of nuclear forces and the detailed structure of nucleons, it is necessary to use particles, especially electrons, which have a de Broglie wavelength (see p. 141) much smaller than the size of the nucleus; for this purpose energies of the order of GeV or higher are required. Thirdly, and most importantly, the production of the strange particles and antiparticles needs energies of many thousands of million electron volts, because of their large rest masses. These are the reasons for the drive in the last two decades to build accelerators of ever increasing energies.

Among the accelerators used for this purpose, the most important is the proton-synchroton. In this machine protons are accelerated in a circular path in the same way as in the cyclotron, but the difficulty due to the relativistic increase of mass is overcome by changing the magnetic field, which is gradually increased from zero to its full value as the particle gains in energy; there is also a corresponding change of the frequency of the accelerating potential. If one can start with particles which have already been accelerated to an energy of 10–50 MeV, for example by using a Van de Graaff or linear accelerator, the orbit of the particle can be made almost constant. This means that one can dispense with the bulky and costly solid magnet and use instead a much cheaper magnet in the form of a ring.

Even so, the weight of the magnet in such machines is colossal. For example, the 10 GeV accelerator at Dubna in the Soviet Union had a magnet weighing 36 000 tons. However, by employing a magnet based on a different principle, the so-called alternating gradient magnet, it became possible to reduce considerably the gap between the poles and thus decrease the weight. The proton-synchrotron at C.E.R.N. (European Organization for Nuclear Research) in Geneva, which came into operation in 1959 and delivers protons of an energy of 28 000 million electron-volts, has a magnet which contains only 3 400 tons of steel; the diameter of the orbit of the protons is 200 metres. This accelerator delivers a pulse of protons every three seconds, each pulse consisting of about 10^{12} particles. Its cost was about ten million pounds. A still larger proton-synchrotron is working in Serpukhov, U.S.S.R. Its magnet has a diameter of 472 metres and it produces protons of energy 76 GeV. At this energy it became possible to produce not only antiprotons and antideuterons but also antihelium-3. The largest proton-synchrotron at the present time is in Batavia (near Chicago) in the U.S.A. It has a diameter of 2 km and produces protons of an energy 200 GeV. In Great Britain a proton-synchrotron producing protons of an energy of 7 GeV is working at the Rutherford National Laboratory for Nuclear Research.

As already mentioned, high energy electrons are particularly useful for the exploration of the structure of nucleons, and several types of electron accelerator have been developed. One of these, the betatron, is in fact an electric transformer in which the secondary coil is replaced by a stream of electrons revolving round the core. Much higher energies are achieved in the synchrotron, in which the acceleration of electrons is achieved in a fashion similar to that in the cyclotron; the frequency of the accelerating voltage is kept constant but the magnetic field is varied so as to maintain a constant orbit of the electrons. Beams of electrons of an energy of 12 GeV have been obtained in such a synchrotron.

A limit to the energy which can be reached in accelerators in which the particles move in a circular orbit is imposed by the loss of energy through radiation. According to the laws of classical electrodynamics, a charged particle which moves on a curved orbit radiates energy in the form of electromagnetic waves. The amount of energy lost increases very rapidly with the kinetic energy of the particle and this is particularly serious for a light particle such as the electron. It is largely because of this, the so-called synchrotron effect, that the acceleration of electrons to very high energies is done by means of a different device, the linear accelerator.

As the name implies, in such an accelerator, the particles move on straight lines and the acceleration is usually achieved by means of a travelling radio-wave set up in a wave guide at a very high frequency (several thousand megaherz). One such linear accelerator, 3 km long, is installed at Stanford University, U.S.A. and produces electrons of energy 20 GeV.

Despite the advances already made in the production of particles of very high energy, the end is not yet in sight. As the energy of the particles is increased, new phenomena are being discovered which make it desirable to have particles of still higher energy, and accelerators to meet this demand are being built. At C.E.R.N. a proton-synchrotron is being constructed to produce protons of an energy 400 GeV; it is hoped that this machine will eventually give protons of 1 000 GeV by using super-conducting magnets, which produce much stronger fields (see p. 155). The Batavia accelerator, too, is expected to reach an energy of 1 000 GeV by using super-conducting magnets. And already plans are being prepared for a 2 000 GeV accelerator.

Another way of increasing the effective energy of the particles is by using colliding beams. Ordinarily, when a fast moving beam of particles hits a stationary target, only a small proportion of the energy is utilized in the nuclear interaction; a very much greater energy is available when two beams of particles, travelling in opposite directions, are made to collide. For this purpose, use is made of the fact that it is possible to store a proton beam in a ring-shaped tube placed in a magnetic field; proton currents of 20 amperes can be made to circulate in such a ring for a day without any significant loss. If two such rings are provided at the exit of a proton-synchrotron, two proton beams travelling in opposite directions can be stored, and made to collide at the appropriate moment. Such a system of Intersecting Storage Rings has been set up at C.E.R.N.; with the energy of protons available from the proton-synchrotron, the result of the collision between the two beams is equivalent to a single proton beam hitting a stationary target at an energy of 1 500 GeV.

The enormous cost of these machines, running into billions of dollars, is becoming a strain even for the super-powers, and it is likely that further progress in high energy nuclear physics will depend on a truly international collaboration, with the pooling of the financial, as well as scientific resources of many nations.

Neutrons as Projectiles

We have so far not mentioned a projectile which does not require high energies, namely the neutron. The employment of neutrons to bombard nuclei has very great advantages. First of all, since a neutron does not ionize the atoms through which it passes, it does not lose any energy in interaction with electrons. It can only interact with a nucleus and consequently every neutron is bound eventually to produce a nuclear reaction. Secondly, as the neutron has no electrical charge it does not experience any repulsive force in approaching the nucleus. Fig. 15 shows the type of potential field around the nucleus as 'seen' by a neutron. Instead of a barrier there is a flat potential until the neutron reaches the nucleus when it is immediately captured and falls into the well. Thus, one does not need neutrons of high energy in order to produce nuclear reactions. In fact, slow neutrons are in this respect much more efficient than fast ones, simply because the lower the speed of the neutron the longer it remains within the sphere of influence of the nucleus and consequently the greater its chance of being captured by it. The slowing down of neutrons can be achieved easily by causing fast neutrons to collide with atoms of a light element, preferably with hydrogen. As already explained, at each collision with a proton the neutron loses a large fraction of its energy and so, after a number of collisions, the kinetic energy of the neutron is reduced to the

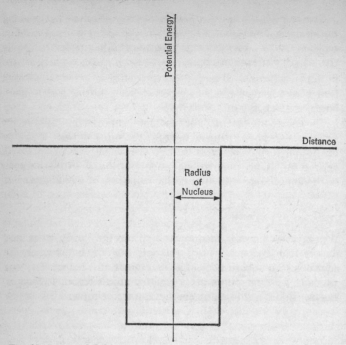

Fig. 15. Potential field around a nucleus as seen by a neutron

energy of thermal agitation with which molecules of the substance move about in a given temperature. Such thermal neutrons, as they are called, have a very high probability of being captured; the cross-section for this process sometimes being of the order of several thousand barns, i.e. several thousand times greater than the actual geometrical cross-section of the nucleus. A very large number of disintegrations can thus be produced by neutrons and many radioactive isotopes have been manufactured in this way. A typical example of a 'capture' process is the production of radioactive sodium from ordinary sodium, according to the equation

$$^{23}Na + {}^1n = {}^{24}Na.$$

The energy released in this reaction (the binding energy of the neutron) is emitted in the form of γ-rays.

The reader may ask, if neutrons are so effective in producing nuclear disintegrations, why do we bother to build the huge machines to accelerate charged particles? The answer to this is that, first of all, the reactions produced by charged particles are different from those produced by neutrons, and we are interested in studying all kinds of disintegrations; secondly, because there are no large sources of free neutrons available in nature. Very soon after a neutron is produced in a nuclear reaction it attaches itself to a nucleus met on its way and ceases to exist as a free particle. The only way to obtain neutrons has been by means of nuclear reactions, such as the bombardment of beryllium with a-particles, or of deuterium with deuterons, in which neutrons are knocked out from nuclei. The problem was, therefore, brought back to disintegrations produced by charged particles.

The situation was radically changed early in 1939 when a new nuclear process was discovered which not only provided a new and extremely rich source of neutrons, but has transformed the whole field of nuclear physics from an academic subject into one carrying the most important practical implications. This process, nuclear fission, will be discussed in the next chapter.

CHAPTER 4

Energy from the Atom

Nuclear Fission

The disintegration experiments described in the previous chapter, while of very great importance as the only means of providing information about the structure of the nucleus, are of little practical value, chiefly because of the small scale on which the nuclear reactions can be made to occur. Even with the most efficient cyclotron the number of nuclei that can be changed in an hour is of the order of 10^{18}, while the number of atoms in one gramme of a medium-weight substance is of the order of 10^{22}. It is obvious that for the practical utilization of the energy of the nucleus one must either contrive a source of particles more intense than those produced in the accelerators or create conditions in which a considerable proportion of *all* nuclei of a substance are involved. Both these methods are now available; the first is based on fission and the second on fusion.

We know already that very heavy nuclei tend to be unstable because they contain many protons, and that if such a heavy nucleus should break up into two, a large amount of energy would be released, because the binding energy of medium-weight elements is much larger than of heavy ones. It turns out that such breaking up of a heavy nucleus, or fission as it is called, can be achieved in various ways but chiefly by adding a neutron to the nucleus. The extra energy brought in by the neutron, which in the case of a slow neutron is its binding energy (about 6 MeV) is sufficient to disturb the delicately balanced equilibrium and to cause fission. The result of such fission is that a nucleus like uranium, breaks up into two fragments of nearly equal weight. The fragments may be nuclei of krypton and barium, or of yttrium and iodine, or of any other two elements whose atomic numbers add up together to 92, the atomic number of uranium. It was in fact the chemical identification by Hahn and Strassmann of medium-weight elements produced as a result of the bombardment of uranium with neutrons, that led Otto Frisch and Lise Meitner to interpret this discovery in terms of fission.

One of the chief features of fission is the large release of energy. This follows from the difference in binding energies per nucleon for heavy and medium elements. From the graph of Fig. 8 it is seen that this difference is nearly 0·9 MeV (million electron-volts) per nucleon. Since the isotope of uranium of main interest has 235 nucleons, the total energy released is about 200 MeV. This energy appears initially mainly as the kinetic energy of the two fragments; owing, however, to their very intense ionization the energy is soon dissipated and converted into heat.

The next feature is that both fragments are radioactive. This, too, is to be expected when we recall that in heavy nuclei the ratio of neutrons to protons is greater than in medium nuclei. The fission fragments have, therefore, too many neutrons, and this can be adjusted in the usual way, by a transformation of a neutron into a proton with the emission of a β-ray. In fact, the neutron surplus is so great that, in general, the new nucleus undergoes again a similar decay, and so we have for each pair of fragments a series of radioactive nuclides following each other. Altogether, over 300 radioactive nuclides, covering 37 elements, from zinc to dysprosium, have been identified as fission fragments.

The most important feature, however, is the emission of several fast neutrons at fission. This, too, is a result of the large surplus of neutrons at the breaking up of the heavy nucleus. In the case of uranium-235 the average number of neutrons emitted at each fission is between 2·5 and 3, depending on the energy of the neutrons causing fission. These few neutrons make all the difference, because they open the way for a self-sustained nuclear chain reaction. Each neutron emitted at fission can be made to hit another uranium nucleus, causing further fission, which emits more neutrons, and so on. If, for the sake of argument, we assume that 2 neutrons are available to produce further fission then if we start with 1 neutron we shall have 2 neutrons in the first generation of fissions, 4 neutrons in the next, over 1 million after 20 generations, 10^{12} after 40 generations, 10^{18} after 60 generations, and 10^{24} neutrons after 80 generations. Thus, we see that we need only to allow the chain reaction to proceed long enough to produce any number of neutrons.

Two conditions must be fulfilled to make such a chain reaction possible. First, we must ensure that the neutrons are not captured by nuclei in a way which does not lead to fission. This means that we must have an assembly of pure fissile material without significant quantities of substances which absorb neutrons. Secondly, we must ensure that the neutrons will not be able to escape from the assembly before producing further fission. Since we know how far a neutron will travel between two collisions, we can calculate the minimum amount of material which is required to maintain a chain reaction. This minimum amount is called the critical mass. It is impossible to set up a divergent chain reaction unless an amount of material greater than the critical mass is assembled.

On the other hand, once such an amount is assembled a chain reaction, with a possible catastrophic result, will take place immediately, since with fast neutrons the time between two successive fissions is about 10^{-9} seconds, and so an enormous release of energy may occur in an extremely short time. This raises the question of how to control the reaction.

The chief way to keep the chain reaction under control is by introducing into the fissile material a substance which absorbs neutrons. Depending on how much of this substance is put in, the number of neutrons available for further fission can be made either smaller or greater than one. Only in the latter case would a divergent chain reaction develop. If, therefore, we start off with a large amount of the absorber in the uranium and then gradually withdraw it we can build up the neutron population in it to the desired value. If the amount of absorber is such that after each fission exactly one neutron is available to produce further fission, the reaction will proceed at a steady rate, with the total number of neutrons, and consequently the power produced, remaining constant.

The gradual build-up of the chain reaction is made possible by the fact that not all neutrons are emitted instantaneously upon fission. A small fraction of them, about 0·6 per cent, are emitted with some delay, ranging from a fraction of a second to about one minute after the fission, and it is these delayed neutrons which make the control so much easier. If we approach the critical condition slowly, so as not to have too large an excess of neutrons, the maintenance of the chain reaction is dependent on the delayed neutrons, because the prompt neutrons by themselves would not be sufficient to make a divergent reaction. The multiplication of neutrons will, therefore, proceed slowly with the period of the delayed neutrons, giving plenty of time to make adjustments. In fact, the multiplication time of the reaction can be made as long as several hours. The power level can thus be controlled very easily and the danger of an explosion is practically eliminated.

Nuclear Reactors

An arrangement in which a chain reaction based on fission can be established and its power level controlled is called a nuclear reactor. Most of the reactors already in existence, or in process of being built or planned, are for the purpose of production of power, but some reactors were built for the specific purpose of providing very high neutron fluxes or for the breeding of fissile materials.

Although all reactors are based on the same principle, they may differ in many characteristics: the kind of fissile material they employ and its distribution; the energy of the neutrons causing fission; the type of moderator, if any; the temperature at which the reactor is run; the method of extracting the heat; and, of course, the power output.

The choice of fissile material often determines other properties of the reactor. Natural uranium contains mainly the isotope ^{238}U; only 1 atom in 140 belongs to the lighter isotope ^{235}U. These two isotopes have quite different fission properties. Uranium-238 requires more than the neutron binding energy to break it up, and consequently only neutrons which have a kinetic energy above 1 million electron-volts are capable of producing fission. The neutrons emitted at fission have a wide distribution of energies, with many below 1 MeV; this means that only a fraction of them can be utilized for further fission. Moreover, uranium-238 has a fairly high probability of absorbing neutrons without producing fission, a process which will be discussed in more detail later on. The result of these two effects is that it is impossible to maintain a chain reaction in uranium-238, no matter how much of the material is assembled.

On the other hand, uranium-235 has been found to undergo fission with neutrons of all energies. In fact, thermal neutrons have a very high probability of producing fission. Furthermore, this isotope does not absorb neutrons appreciably. Consequently, it is very easy to set up a chain reaction in uranium-235, and the critical size is very small, of the order of one kilogramme. Since, however, uranium-235 is present in natural uranium in a very low proportion, it has to be separated from the bulk by means of one of the very laborious and costly techniques of isotope separation.

In natural uranium, with a 0·7 per cent concentration of ^{235}U, it is impossible to maintain a chain reaction with fast neutrons, since such neutrons have only a small probability of producing fission in uranium-235. On the other hand, it is possible to achieve this with slow neutrons. Since the neutrons emitted at fission are fast ones, they have to be slowed down before they hit other uranium nuclei. A reactor in which the bulk of the fission is caused by slow neutrons is known as a thermal reactor. The slowing-down process is called moderation and can be achieved by making the neutrons collide with atoms of light elements. The best element would be hydrogen, but it also absorbs neutrons, and the combined loss of neutrons by hydrogen and uranium-238 is so great that the reaction is inhibited. However, hydrogen can be used as a moderator, usually in the form of natural water, if the uranium is enriched in the 235 isotope.

If natural uranium is to be used as a fuel other materials must be employed as moderators. The best is deuterium, in the form of heavy water. However, deuterium is fairly expensive, and for this reason most thermal reactors use carbon in the form of graphite as the moderator. Such a reactor consists of an assembly of graphite bricks in which uranium rods are inserted at intervals, forming a lattice. The neutrons emitted from the uranium pass through the graphite, where they are slowed down, and as such have a much higher probability of causing fission in the uranium rods. Owing to the low proportion of the fissile material and to the relative inefficiency of carbon as a moderator the critical size is much greater, and may be of the order of 10 tons or more of natural uranium. The total weight of the uranium and graphite may be about 1 000 tons, but the main weight in the reactor assembly is of the concrete of the surrounding walls which are necessary to provide shielding from the intense neutron and γ-radiation issuing from the reactor.

Although uranium-235 is the only naturally occurring fissile material in which a chain reaction can be set up, there are several other artificially produced materials which can serve the same purpose. One of these is plutonium-239, a nuclide of atomic number 94, which is obtained from uranium-238 as a result of the latter capturing a neutron. Uranium-238 is then transformed into uranium-239, which is a radioactive nuclide and decays by the emission of a β-ray into neptunium-239. This too is radioactive, and after another β-emission is transformed into plutonium-239.

Since plutonium does not exist in nature it has to be produced, atom by atom, by means of this process. To carry this out on a practical scale one would need an enormous number of neutrons, which can only be provided in a nuclear reactor. In fact, the first nuclear reactors, which were based on natural uranium, were built for the sole purpose of manufacturing plutonium. Once having produced a sufficient amount of plutonium this can then be used by itself as a fissile material in the same way as uranium-235. Thus, a nuclear reactor can be used not only to generate power, but also to breed new nuclear fuel by utilizing some of the neutrons in the reactor to convert ^{238}U into ^{239}Pu.

Another possibility is to utilize thorium. If thorium-232 is bombarded with neutrons it is converted into the isotope thorium-233, which decays into protactinium-233, and this in turn into uranium-233. Uranium-233 is a fissile material with properties similar to those of uranium-235 or plutonium-239.

Nuclear Power Production

Although nuclear reactors have been used for the production of energy since 1955, there is still no general agreement about the optimum design, particularly concerning the moderator, the material used for cooling, and the method of extracting the heat from the reactor.

In all power reactors the energy released from the nucleus at fission is ultimately converted into electricity. In theory one could extract the energy directly in the form of electricity by utilizing the fact that the fission fragments carry electrical charges. In practice this has not been possible as yet, and the production of energy is achieved in a roundabout way: the fission fragments lose their energy by setting in motion the atoms through which they pass, in other words in the form of heat. This heat is then extracted and converted into electrical energy by means of a conventional turbine. The heat is extracted by passing through the fissile fuel rods a cooling medium (coolant), which may be either gaseous or liquid; among the latter, water, organic compounds and liquid metals (sodium or potassium) have been used. The coolant is then brought into thermal contact with water to produce steam to drive the turbine. The efficiency of converting to electricity depends on the temperature of the steam in the turbine; this is one of the deciding factors in the economy of nuclear power production.

Practically all power reactors operating at the present time are thermal reactors, based either on natural uranium or on uranium enriched in the 235 isotope, and employing a moderator to slow down the neutrons to thermal energies.

In the United Kingdom, where electrical power from nuclear reactors has been fed into the national grid since 1956, the first series of reactors used natural uranium metal as the fuel, and graphite as the moderator. The extraction of heat was achieved by passing through the uranium rods a gas, carbon dioxide, at a pressure of 7 atmospheres. Because of these characteristics these reactors are known as Gas-Cooled Reactors. As shown in Fig. 16, the carbon dioxide is circulated in a closed system; this is necessary since the gas carries with it some of the radioactive fission

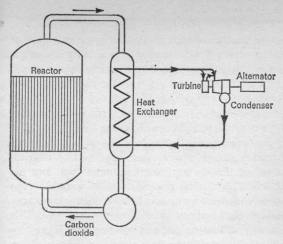

Fig. 16. Extraction of heat from nuclear reactor

products. The heat is utilized in the heat extractor in which water is brought into thermal contact with the hot carbon dioxide and converted into steam which is then used to drive a turbine. The temperature of the steam entering the turbine is about 600 kelvin.*

The second series of reactors, which came into operation in 1965, is called Advanced Gas-Cooled Reactors (A.G.R.). It differs from the first in that the uranium has been enriched in the 235 isotope to double its natural concentration (1·4 per cent). The method of sheathing the fuel rods of uranium oxide has also been changed. These changes, together with some other improvements, have made it possible for the reactor to work at much higher temperatures, e.g. steam temperatures of 838 K, with a consequent greatly increased efficiency in converting heat into electricity; this efficiency is now 42 per cent. In parallel with this the total power output of the reactor has also been increased; while in the first series the average power output per reactor was about 200 megawatts of electricity, in the A.G.R. reactors the output is

* The kelvin (K) is the unit of absolute temperature: 0 kelvin is −273·16° Celsius; the magnitude of the unit is the same in both scales.

625 megawatts. By 1975 the total nuclear generating capacity in the U.K. will amount to 11 000 megawatts of electrical power.

Other types of reactors which are being developed are the Steam-Generating Heavy-Water Reactor (S.G.H.W.R.), in which the moderator is heavy water; the High-Temperature Reactor (H.T.R.), which uses helium as the coolant; the Pressurized-Water Reactor (P.W.R.) and Boiling-Water Reactor (B.W.R.), in which ordinary water is used both as the moderator and coolant. The last two types are developed mainly in the United States.

From the point of view of economy, large reactors are much more advantageous and it is envisaged that reactors with an electricity output of 1 000 megawatts or higher will become standard. Each such reactor will require a load of some 650 tons of uranium oxide enriched in uranium-235. The yearly consumption of uranium would be about 200 tons.

With the rapidly increasing demand for electrical energy all over the world, there is likely to be a very fast growth of the nuclear power industry, and it is estimated that by the year 2000 over two million megawatts of electricity will be generated in nuclear reactors. This would entail the consumption of some 400 000 tons of uranium every year, which is a high proportion of the known world reserves of economically mined uranium ore. The rapid exhaustion of uranium-235 can be forestalled by the development of reactors which utilize the other 'fertile' materials, uranium-238 and thorium-232, and convert them into the fissile nuclides, plutonium-239 and uranium-233. In fact, the whole future of nuclear energy based on fission is dependent on the development of so-called breeder reactors in which new nuclear fuel is being produced at the same time as the original fuel is being burned up. By means of such breeder reactors the total reserves of nuclear fuel can be increased by a factor of about 100.

For high breeding efficiency all neutrons emitted at fission have to be utilized, and therefore breeder reactors do not use any moderator, and the chain reaction is based on fast neutrons. Several prototype fast breeder reactors have been developed, the most advanced is at Dounreay. It has a compact core of 900 kg of PuO_2, and with some 3 000 kg of UO_2 around it. The coolant is liquid sodium at a temperature of up to 873 K. The power output is 250 megawatts of electricity. The first commercial fast breeder reactors will produce 1 300 megawatts of electricity and are expected to come into operation at the end of the 1970s.

The fantastic quantities of radioactive materials produced in reactors has caused considerable concern about the harmful effects to living organisms in case of an accident to the reactor in which radioactivity may be released. Even without an accident some radioactivity may escape during the very complex procedure of separating the accumulated fissile materials from the fuel rods, which has to be done from time to time to ensure proper running of the reactor. The storage of the huge quantities of radioactivity, some of which have long half-lives of the order of 30 years, will also become a problem of increasing difficulty as time goes on.

The Atom Bomb

In the discussion of nuclear reactors emphasis was laid on the conditions which ensure steady operation of the reactor, so that the power output will be constant. Quite opposite are the requirements for a weapon. In this case it is desired to release a large amount of energy in the shortest possible time. It is obvious that fast neutrons must be used for this purpose, because the slowing-down time of a neutron is of the order of 10^{-4} seconds, which means that it may take about one hundredth of a second to produce some 80 generations of neutrons. This is far too long a time to make an effective weapon. With fast neutrons, however, 80 generations of neutrons could be developed in less than a millionth of a second. This is the reason why an atom bomb needs either pure uranium-235 or plutonium-239, and it was for this purpose that the first nuclear reactors as well as the big isotope separation plants were developed during the war.

In principle, the mechanism of exploding the atom bomb is quite simple. One starts with two pieces of fissile material, each a little smaller than the critical size, and brings them together quickly at the moment when the bomb is to be exploded. The emphasis is on quick assembly, as otherwise the heat developed in the earlier fissions may disperse the system before a large number

89

of fissions have had time to build up. A rapid assembly may be achieved by shooting the two parts of fissile material together by means of ordinary explosives. If, say, 10^{24} fissions are developed, the heat produced will be about 10^{14} joules and the explosive power equivalent to about 20 000 tons of T.N.T. This was said to be the explosive power of the first atom bomb, dropped over Hiroshima in 1945, which was based on the above principle.

The second atom bomb which was exploded over Nagasaki was based on an implosion mechanism. The fissile material is arranged in the form of a number of spherical segments, each smaller than the critical size. On the outer edge of each segment is placed an ordinary explosive, shaped in such a way that when detonated it drives the segment towards the centre. All the shaped charges are detonated simultaneously and the force of the implosion compresses the fissile material, considerably increasing its density. At a higher density the critical mass is smaller and the time between successive fissions is reduced. Both these factors contribute towards the building up of the chain reaction in a much shorter time and the releasing of a much larger amount of energy than with an ordinary assembly. It was stated that by the use of an improved implosion mechanism the explosive power of the bomb was subsequently increased twenty-five-fold, making it thus equivalent to about 500 000 tons of T.N.T. (0·5 megatons). This seems to be the practical limit of the explosive power of an atom bomb.

At the other extreme, the implosion mechanism has also made it easier to produce bombs of relatively low explosive power which are intended as 'tactical' weapons. In every fission bomb an amount of material of about the critical size is wasted since the chain reaction stops as soon as the remaining fissile material is reduced to below critical size. In an implosion bomb the critical size may be several times smaller than under ordinary conditions, thus making it more economical. By using the appropriate amount of fissile material the explosive power of the bomb can be reduced to the desired level. It has been reported that atom bombs with an explosive power of less than a ton of T.N.T. have been developed, which can be fired from an ordinary gun.

Thermonuclear Reactions

The fission process has made possible for the first time the release of the energy in the nucleus on a practical scale. There is, however, yet another way in which this could be achieved. From the binding energy curve (Fig. 8) it is clear that energy would be released at the fusion of light elements into heavier ones. In particular, we have found that helium is a very stable nucleus; if, therefore, hydrogen could be converted into helium, a large amount of energy would be released.

The previous discussions have indicated the difficulties which would be encountered in attempting to produce large-scale nuclear disintegrations by means of charged particles. First, the majority of such particles would lose their energy on ionizing the atoms through which they pass. Secondly, the particles would have to possess a sufficiently high energy to penetrate the potential barrier of the nucleus, and we have seen that it would be very difficult to produce a really large number of such projectiles. At first sight, therefore, it would appear that the maintenance of a nuclear reaction by means of charged particles is impossible on a large scale. The situation is, however, quite different if we consider a system at a very high temperature. At a temperature of the order of many millions of degrees, the atoms are stripped of their electrons, and all matter consists of a mixture of bare nuclei and free electrons moving about with the speed of thermal agitation. There is, therefore, no loss of energy at collisions with electrons, and the nuclei have a good chance to collide with each other. If the temperature were so high that the nuclei had an energy greater than the potential barrier, every such collision would give rise to a disintegration. We know, however, that the energy of a projectile need not be as great as the potential barrier in order to produce a disintegration. The wave properties of the particle enable it to penetrate through the potential wall even if its energy is quite low. Indeed, it has been found that protons of an energy as low as a few thousand electron-volts can produce disintegrations in light elements. Naturally, the lower the energy the smaller the probability of penetrating the barrier. Fig. 17 shows a typical excitation function, or probability of producing

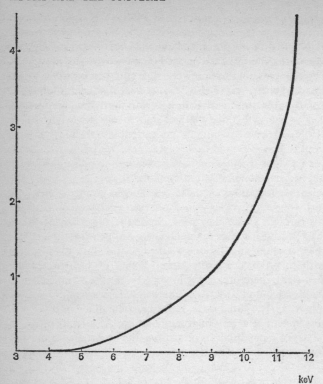

Fig. 17. Excitation function at low energies

a disintegration as a function of the energy of the bombarding particle. It is seen how very rapidly the probability goes up with an increase in energy. If, however, we have a large number of projectiles which do not lose energy in any other way, and if we can afford to wait long enough, then nuclear disintegrations are bound to take place even at low energies.

These considerations show that it should be possible to produce large-scale nuclear disintegrations if a substance is heated to a temperature at which the particles have an energy of a few thousand electron-volts. An energy of about 10 000 eV (10 keV) corresponds to a temperature of 116 million kelvin, but a somewhat lower temperature would still do. This is so because at a given temperature not all particles move with the same energy. If we were to measure the energy of individual particles at a given temperature we should find a very wide distribution, of the type shown on Fig. 18. This distribution of energies is for a temperature of about 20 million kelvin. It is seen that while the *average*

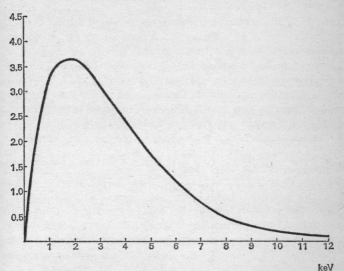

keV

Fig. 18. Distribution of energies of particles at a temperature of 20 million kelvin.

energy of the particles at this temperature is about 2 keV, there are many particles which have an energy between 5 and 10 keV. Thus, we conclude that if hydrogen could be heated to a temperature of about 20 million kelvin, and maintained at this temperature for a sufficiently long time, a large-scale thermonuclear reaction could take place.

Nuclear Reactions in the Sun and Stars

The conditions just described are fulfilled in the sun, which is composed mainly of hydrogen and whose interior temperature is of the order of 20 million kelvin. We may, therefore, expect that thermonuclear reactions take place in the sun. In fact, as will be shown in Chapter 9, it is impossible to account for the magnitude of the energy emitted from the sun and for the steady rate of this emission over millions of years by any other source of energy than nuclear. We have, of course, no direct proof as to the type of thermonuclear reaction occurring there, but from our knowledge of the Q-values and cross-sections for various reactions, and taking into account the temperature, size and density of matter in the sun, it is fairly easy to deduce which of the known nuclear reactions might be responsible for the production of energy in the sun. Such calculations were first made by Bethe in 1938 and they have provided us with a fairly good basis for understanding the nuclear processes going on in the sun. The two chief processes which we now believe are responsible for the sun's energy are the hydrogen chain and the carbon cycle.

The hydrogen chain is represented by the following formulae:

$$^1H + {}^1H = {}^2D + e^+,$$
$$^2D + {}^1H = {}^3He,$$
$$^3He + {}^3He = {}^4He + {}^1H + {}^1H.$$

We start off with two protons colliding with each other. An assembly of two protons is, of course, unstable, owing to the electrostatic repulsion between them, but it may become a stable structure if one of the protons changes into a neutron with the emission of a positron. Thus, the result of this encounter is the production of a nuclide containing one proton and one neutron, in other words, a deuteron, the nucleus of heavy hydrogen. In the next stage, another proton collides with the deuteron to produce the nucleus of helium-3, containing 2 protons and 1 neutron. In the final stage two such helium nuclei, formed in two different collisions, meet each other. They combine to form a nucleus of helium-4, and 2 protons. The final balance, therefore, is that we have put into the reaction 6 protons and obtained an a-particle, 2 protons and 2 positrons. Effectively, therefore, it is a fusion of 4 protons into an a-particle. We know already that the a-particle is a very stable structure and that its binding energy is about 28 MeV.

The other process, which is called the carbon cycle, takes place in four stages, which are represented in the following formulae:

$$^{12}C + {}^{1}H = {}^{13}N \rightarrow {}^{13}C + e^{+},$$
$$^{13}C + {}^{1}H = {}^{14}N,$$
$$^{14}N + {}^{1}H = {}^{15}O \rightarrow {}^{15}N + e^{+},$$
$$^{15}N + {}^{1}H = {}^{12}C + {}^{4}He.$$

In the first stage a proton hits a carbon nucleus. The result of this reaction is the formation of a nucleus of nitrogen-13 containing 7 protons and 6 neutrons. This nucleus, however, is unstable, as it contains too many protons, and consequently one of the protons changes into a neutron, with the emission of a positron, resulting in the formation of a nucleus of carbon-13, which has 6 protons and 7 neutrons. In the next stage this carbon nucleus is hit by another proton, which combines with it to form the nucleus of nitrogen-14 containing 7 protons and 7 neutrons.

In the third stage this nitrogen nucleus is hit by a third proton. The nucleus formed, oxygen-15, contains 8 protons and 7 neutrons and is again unstable, having too many protons. One of the protons changes into a neutron with the emission of a positron, resulting in the formation of nitrogen-15. Finally, in the last stage, a fourth proton hits this nucleus nitrogen-15, breaking it up into the nuclei of carbon-12 and helium. Thus we have got back the carbon nucleus with which we started the reaction. This means that carbon itself is not being used up in this process but only acts as a catalyst. The final balance is that we have put in 4 protons and obtained 1 a-particle and 2 positrons.

We see, therefore, that in both these processes, the hydrogen chain and the carbon cycle, we deal ultimately with the same reaction, with the fusion of hydrogen into helium, with the simultaneous emission of energy. Which of these processes takes place predominantly depends on various other conditions, particularly on the size of the star. It is nowadays believed that in the sun the hydrogen chain is the predominant reaction, while in larger stars the carbon cycle is the chief source of energy.

At the temperature of the sun both the hydrogen chain and the carbon cycle are very slow processes. Apart from the fact that several stages of the reaction involve radioactive transformations which are slow in time, the main delay is due to the small probability of penetrating the barriers at collision. Thus, the period of time for the whole carbon cycle to take place is calculated to be millions of years.

The reason why the energy in most of the stars comes from the burning of hydrogen is because hydrogen is the main constituent of stars; since it requires the lowest temperature for a thermonuclear reaction, only hydrogen fusion takes place. Should, however, the supply of hydrogen be exhausted, then other elements, in the first place helium, may take over to produce thermonuclear reactions. The exhaustion of hydrogen would cause the star to contract through gravitation and thus increase its temperature until about 10^8 kelvin, when thermonuclear reactions in helium may occur. As a result of burning of helium, carbon may be produced, and after helium has been exhausted a further contraction may raise the temperature still further, to about 6×10^8 kelvin, when carbon may be ignited, and so on. There is reason to believe that some stars have already reached those stages, but the majority of stars still burn hydrogen as this is the most abundant element in the universe.

The Hydrogen Bomb

The possibility of setting up a thermonuclear reaction on the earth was considered unrealistic until 1945, since the highest temperatures which could at that time be achieved in the laboratory were much less than 100 000 kelvin, while for thermonuclear reactions a temperature of the order of tens of millions of degrees is necessary. The situation changed, however, after the development of the atom bomb based on fission. At the instant of the explosion the temperature reaches several million degrees, and although this lasts only an extremely short time it may be sufficient to initiate a fusion reaction. By its very nature such a reaction could only be utilized as an explosive, and such an arrangement is known as the hydrogen bomb.

Despite its name it is clear that ordinary hydrogen cannot be used as an explosive. We have already seen that the solar reactions, the hydrogen chain and the carbon cycle, are very slow. Any process which depends on a β-decay, the transformation of a proton into a neutron, is far too slow to be suitable for a thermonuclear explosion. There are, however, isotopes of hydrogen which appear to be more suitable for thermonuclear reactions. One is the so-called d-d reaction in which two deuterons are made to collide. The result of the collision can be either of the two following processes:

$$^2D + {}^2D = {}^3He + {}^1n$$
$$^2D + {}^2D = {}^3T + {}^1H.$$

The Q-values for these processes are 3·27 and 4·03 MeV respectively. The first reaction has already been mentioned (Chapter 3) as a source of neutrons.

The other reaction, the d-t process, is based on a third isotope of hydrogen, called tritium (symbol T), the nucleus of which, the triton, consists of 1 proton and 2 neutrons. When a deuteron and a triton meet, they are converted into an α-particle and a neutron according to the formula

$$^3T + {}^2D = {}^4He + {}^1n,$$

with a Q-value of 17·59 MeV.

Because of its higher Q-value the tritium reaction will produce much more heat for a given number of collisions than the d-d process. Much more important, however, is the fact that at relatively low temperatures the tritium reaction proceeds much faster than the deuterium one. A comparison of the probabilities for the occurrences of the various reactions shows that at these temperatures the tritium reaction yields several hundred times more heat per gramme of substance burned than the deuterium reaction. It is believed that the short time during which the necessary high temperature is maintained in the atom bomb is not sufficient to start a reaction in deuterium alone, and that consequently a mixture of tritium with deuterium is necessary. If this were true, a hydrogen bomb would be prohibitively expensive owing to the high cost of tritium. This nuclide does not exist in nature, and can be obtained only through a nuclear reaction. The best process seems to be the bombardment of lithium with neutrons according to the formula

$$^6\text{Li} + {}^1\text{n} = {}^4\text{He} + {}^3\text{T}.$$

To make tritium on a large scale one would have to put the lithium into a nuclear reactor where it would be bombarded with neutrons. This process of manufacture of tritium is, therefore, somewhat similar to that of making plutonium, except that one would need 80 times more neutrons to make tritium than to make the same weight of plutonium. Moreover, while plutonium has a very long half-life (about 24 000 years) the half-life of tritium is only 12·3 years, and so it would be very wasteful to store the tritium.

Apart from the high cost of tritium, the deuterium and tritium would have to be used in liquid form in order to obtain a sufficiently high density. The need of a liquefying system would make such a 'device' even more bulky.

It is believed that a solution to this difficulty was found by making tritium on the spot, at the instant of the explosion, instead of in a nuclear reactor. For this purpose a compound lithium-6 deuteride (^6LiD) is employed. If this substance is put round an atom bomb assembly, then at the instant of the explosion the neutrons emitted at fission bombard lithium-6 nuclei, producing tritium. At the high temperature of the explosion the tritium combines with the deuterium to start a thermonuclear reaction.

Many such hydrogen bombs have been exploded for test purposes, and in one of them an explosive power of about 60 million tons of T.N.T., or 3 000 times greater than the Hiroshima bomb, has been achieved. It is believed that much of the explosive power is derived in these bombs not from the thermonuclear reaction but from fission induced by fast neutrons in an outer shell of uranium-238. An outer shell of a heavy metal is required to prevent too early dispersal of the fusion material; by making this shell of ^{238}U an additional source of energy is obtained. It will be noted that in most thermonuclear reactions neutrons are emitted, and these neutrons are fast enough to produce fission in ^{238}U. There are, therefore, three steps in such a hydrogen bomb: (i) fission in the atom bomb core giving rise to a high temperature and to the emission of neutrons; (ii) fusion in the lithium deuteride by the heat generated; (iii) fission in the outer shell by the fusion neutrons. Since about one half of the explosive power in such a fission-fusion-fission bomb comes from fission processes, an extremely large amount of radioactivity is produced; for this reason such bombs are called 'dirty'. In 'clean' bombs, which have also been developed, most of the energy (about 95 per cent) is derived from thermonuclear reactions. Moreover, it may become possible to produce the high temperature by other means, e.g. by using a laser, and thus avoid the fission products altogether.

Controlled Thermonuclear Reactors

The realization that the energy radiated by the sun is derived from thermonuclear reactions and the development of the hydrogen bomb, gave impetus to large-scale research into the possibilities of the peaceful utilization of fusion reactions. It will be clear from the discussion earlier in this chapter that a large-scale release of nuclear energy based on fusion could be expected only by means of a thermonuclear process. For this purpose it is necessary first to heat up the reacting material at a sufficiently high density to a very high temperature, of the order of many millions of degrees, and then to maintain this temperature long enough for a sufficient number of interactions to take place to sustain the process spontaneously. To be of any practical value, the amount of energy released must be sufficiently large, yet the rate of its release must be controllable. Finally, it should be possible to extract the energy, yet the reaction must be contained within walls which will neither burn up nor quench the reaction. Each of these problems presents enormous but surmountable technological difficulties.

A substance heated to a temperature of millions of degrees is neither solid nor liquid; it is not even a gas in the ordinary sense. The atoms of the material are unable to retain all electrons in the orbits around the nuclei; some or all electrons are torn off and drift freely through the gas. The residual atoms are thus

99

positive ions which also move about with great speed. Such a mixture of ions and electrons is called a plasma.

Surprisingly little is still known about the properties of this fourth state of matter, considering that most of the matter in the universe, e.g., the sun and the stars, is in this state. However, from a knowledge of the properties of various thermonuclear reactions it has been possible to calculate the conditions necessary for a plasma to yield nuclear energy. These calculations have shown that if pure deuterium were used a temperature of about 200 million kelvin would be necessary, while for a mixture of deuterium with tritium about 30 million degrees would suffice. Further, the density of the plasma must be about 10^{15} particles per cm³. At a normal temperature this would correspond to a pressure of only 0·01 mm – a fairly good vacuum – but at the high temperatures of the plasma the gas pressure would be about 30 atmospheres. Under such conditions the power output would be of the same order of magnitude as from a fission power reactor, provided the plasma could be maintained at the required temperature for at least 10 seconds in pure deuterium, or 0·1 second in a D-T mixture.

The attainment of a very high temperature for a very short time is in principle possible. Since the plasma is a very good conductor of electricity one can pass through it a heavy current which will heat it up. In practice this can be achieved by making the plasma the secondary winding of a pulsed transformer. Electrical energy is supplied through the primary winding of the transformer by the sudden discharge of a large bank of condensers. Currents of the order of a million amperes can thus be achieved.

The containment of the hot plasma is best achieved by means of magnetic fields. In one method, the electric current from the plasma itself provides the required magnetic field. This is the so-called pinch effect, in which the magnetic field produced by the current in the plasma constrains it to a thin ribbon along the axis of the tube containing the plasma, thus keeping it away from the walls. Another technique is based on so-called magnetic mirrors, in which external magnetic fields are arranged so as to reflect the particles of the plasma back to the centre of the vessel, thus preventing them from reaching the walls.

Many intricate and costly machines, based on the above principles, but using different devices to achieve the desired aim, have been tested since 1955, but so far none has been successful. The nearest approach to the conditions for a sustained nuclear reaction was in a Soviet machine called Tokamak which is based on the pinch effect. In this device a temperature of 5 million kelvin was maintained for 20 milliseconds in a plasma of density 5×10^{13} ions/cm³. This still leaves a long way to go, but this achievement is considered by experts to be very encouraging.

A great deal of fundamental research on plasma properties has been carried out in recent years and much more will have to be done before the problem of the practical utilization of fusion reactions is solved. This research is very extensive and costly but it is being undertaken because of the enormous practical advantages which it may yield: (i) thermonuclear power will use a fuel which is universally available and is plentiful; the deuterium contained in the oceans would satisfy the world's energy requirements for millions of years; (ii) the fuel would be very cheap; 1 gramme of heavy water, costing at present about 10p, could provide the same amount of heat as 10 tons of coal; (iii) the release of energy in such reactions is not accompanied by the production of large quantities of radioactive materials as is the case in fission reactions. Judging from the rate of progress of science and technology, there can be no doubt that the problem of controlled thermonuclear power will be solved in the foreseeable future.

CHAPTER 5

Matter and its Properties

Inter-Atomic Forces

So far we have dealt only with single atoms, the elementary particles of which they are made, and the changes in structure which can occur in nuclear reactions and radioactivity. We have hardly mentioned the arrangement of the electrons in the outer parts of the atom, which determine how the atoms appear to each other. The importance of nuclear reactions today has tended to make us forget that only the 'minor' nuclear reactions associated with natural and artificial radioactivity were known before 1939, that is, before nuclear fission was discovered. Until that year the word 'reaction' would probably have been assumed to refer to chemical reactions, in which only the outermost parts of atoms are affected. In this chapter we deal with the interactions between whole atoms, which give rise to the familiar phenomena of chemistry and to the existence and behaviour of matter in bulk.

Perhaps the best starting-point in this discussion will be the assertion that every atom exerts a force upon every other atom. The details and the magnitude of the force vary as between one type of atom and another, but in general the force is always a force of attraction when the atoms are at a distance apart greater than their normal diameters, changing to a force of repulsion if the atoms are forced very close together. Thus there will be a tendency for atoms to draw together and 'stick'. This can be very conveniently illustrated in graphical form. In Fig. 19 we see a typical graph relating the potential energy of interaction of two atoms to the distance between their centres. It will be seen that the curve has been drawn so that the energy is zero at large distances. This is clearly a reasonable way of fixing the zero, or reference level, where we are dealing with energies of interaction, because there will be no interaction at very large distances apart. Then at smaller distances the energy adopts increasingly large negative values until the curve eventually reaches a minimum and turns upwards, finally reaching large positive values at very small distances. Two atoms 'held' very close together (where the energy of interaction is positive) would tend to fly apart if released. On the other hand, if they were held at an appreciable distance

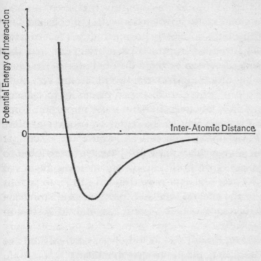

Fig. 19. Potential energy of interaction of two atoms

apart – but not so far apart that there was no perceptible interaction – they would tend to move together. When this curve is examined by the methods of calculus (see Appendix) it turns out that the slope at any point is equal to the net force of attraction. Thus at large distances there is, as already explained, a force of attraction, and at very small distances a force of repulsion. The position at which these cancel each other out, so that there is no net force, will be the normal relative position of the two atoms when stuck together. This is clearly at the minimum – or 'trough' – where the slope is zero.

We must now try to understand something of the origin of these inter-atomic forces, as far as is possible with our limited equipment. Perhaps the easiest kind of force to understand is the repulsive force which comes into play at small distances apart. If there were no such repulsive force, atoms would not have individual existence and would all merge into each other. So the very existence of discrete atoms is a *de facto* demonstration of the existence of repulsion. But we can see a possible origin of this force in the structure of the atom as a positively charged nucleus surrounded by a cloud of negative electrons. If we consider two such atoms approaching so closely together that they overlap greatly (assuming for the purposes of this argument that this does not cause any distortion of the individual atoms) we eventually arrive at a condition in which the two nuclei are very close together, and the two spherical electronic clouds have become nearly coincident. It is obvious that the most important force under these circumstances will be the repulsion of the two positively charged nuclei upon each other, because the electronic clouds will be so thoroughly mixed up that their contributions will act almost equally on both nuclei. At small distances the repulsion of nuclei becomes very powerful and this is, in part at least, the origin of the general repulsive force between atoms.

The attractive forces are of many kinds, and in order to understand them we have to know some details of the outer structure of atoms. It is known, mainly as a result of the detailed study of spectra of all kinds, that the electrons surrounding an atom do not behave as a vague 'cloud' of charge, but occupy, in some way, different zones at varying distances from the atomic nucleus. These zones are called shells, and they can be further divided into sub-shells. Further, there are certain favoured numbers for the occupation by electrons of these shells. Thus the innermost shell cannot hold more than two, the next outer shell not more than eight, leading to larger numbers for shells still further out. The chemical behaviour of atoms is then found to depend on whether the number of their electrons is sufficient to nearly fill, just fill, or overfill a particular one of these shells.

This can be explained by considering as examples three atoms, fluorine, neon, and sodium, which are adjacent in the Periodic Table of elements (Fig. 1), according to their atomic numbers, having respectively 9, 10, and 11 outer electrons. In spite of being so close together in mass and nuclear charge, they are quite different in behaviour. Fluorine is a gas which reacts violently with everything with which it comes into contact – so violently indeed that it is only in recent years that it has been possible to study its chemical properties under controlled conditions. Neon is an 'inert gas', which can hardly be induced to enter into any chemical reaction; for this reason it can be used as a protective atmosphere in experiments at high temperatures. Sodium is a soft solid which reacts violently with water and is a constituent of many common compounds such as sodium chloride (common salt) and sodium bicarbonate (soda).

The clue to their behaviour is given by the case of neon. It has 10 electrons, sufficient to fill two electronic shells with 2 and 8 electrons respectively. It therefore has no tendency to lose or gain an electron, and this is the key to its remarkable chemical inertness. Fluorine, on the other hand, has a tendency under certain circumstances to gain an electron in order to fill a shell, in so doing becoming as a whole negatively charged, or ionized. Similarly, but conversely, sodium has a tendency to lose its extra electron, becoming as a whole positively charged.

Now we can consider the situation when a sodium atom and a fluorine atom find themselves within reasonable distance of each other. They can satisfy their needs to have just filled electronic shells by transferring the 'spare' electron from the sodium atom to the fluorine atom. The atoms have now become positively and negatively charged ions, respectively, and they will exert a purely electric force of attraction upon each other, which may lead to the formation of a molecule of sodium fluoride, or perhaps, if there are very many such pairs of atoms, to a solid lump of this substance.

It is not so easy to explain why, for example, fluorine atoms are hardly ever found singly, but always in pairs, as symbolized by F_2. This is an example of a different kind of attractive force which is of the greatest importance in chemistry. Again, we could

105

not give a true explanation without use of some of the most difficult concepts in modern physics and chemistry (those of quantum mechanics), but we can give a rough explanation, and one which happens to coincide with that offered by chemists before the modern developments were understood. It is that the two fluorine atoms, each having 7 electrons in its outer shell, *share* one of these between them in order to achieve the full complement of 8. This is illustrated in the following 'equation' representing the formation of a molecule of fluorine from two atoms.

$$\overset{\cdot\cdot}{\underset{\cdot\cdot}{:}\text{F}}\cdot \; + \; \cdot\overset{\cdot\cdot}{\underset{\cdot\cdot}{\text{F}}}: \; \to \; \overset{\cdot\cdot\cdot\cdot}{\underset{\cdot\cdot\cdot\cdot}{:}\text{F}}:\text{F}:$$

It can readily be imagined that two atoms will find it increasingly difficult to share more than one electron, for simple geometrical reasons, although there may be no such reasons to prevent one atom sharing several of its electrons among two or more other atoms. Considerations such as these, involving the actual shapes of molecules, the different quantities of energy associated with the various electronic arrangements, and particularly the actual shapes of the orbits in which the electrons are supposed to be localized, make up a large part of modern chemistry when viewed properly in the light of modern concepts rather than in the rough way adopted here.

We must now consider the attraction between atoms of neon or other inert gases (argon, krypton, etc.), and the question of how there may be an attractive force between *molecules* such as F_2 in spite of the fact that the atoms of the molecule have, at least to a first approximation, satisfied their own requirements. There are various kinds of force involved, all much weaker than those already mentioned, and therefore called 'secondary' forces, as compared with the strong primary forces between 'unsatisfied' or unsaturated atoms. The attraction between these atoms or molecules is essentially due to forces between electric dipoles – the electric analogues of ordinary magnets. Two magnets will always attract or repel one another, and if one or both can rotate freely they will adopt a position which makes the force attractive. The characteristic of an ordinary magnet is that there is an apparent separation of magnetic poles (the analogues of electric charges) to the two ends. Although, in an electrically neutral atom such as the neon atom there is no separation of electric charge, there are temporary fluctuations in the distribution of charge which give rise to temporary dipoles. One such dipole will act upon a neighbouring atom in much the same way as one magnet causes another to rotate, so that there is a temporary attraction. It will always be an attraction, whatever the direction of the original dipole, and there is, in fact, a steady though relatively weak attraction between any such pair of atoms or between atoms and molecules, or between pairs of molecules. In some molecules there may be permanent dipoles arising from shape and the actual separation of charge during the formation of the molecule, and here the secondary forces of attraction are somewhat stronger. The difference between the effects of the temporary dipoles in non-polar molecules, as they are called, and of the permanent dipoles in polar molecules, can be noted in many familiar phenomena. For example, benzene and water do not mix. Liquid benzene exists because of non-polar attractive forces, liquid water because of polar forces; the attraction of the water molecules for each other is so much greater than, say, of a water molecule for a benzene molecule, that benzene molecules are, so to speak, 'squeezed out' of the neighbourhood of water. Attempts to mix the two liquids by violent shaking result only in a collection of separate droplets of benzene and water.

Finally, there is an attraction between metallic atoms, which differs from those already discussed. (It cannot be explained in simple terms, but only with the help of modern quantum theory. The reader may perhaps be asked to take this on trust, with the assurance that he has many allies even among physicists and chemists in this.) A characteristic example is provided by metallic sodium. Here the 'spare' electrons, one from each atom, appear to be shared among all the sodium atoms, and act to bind them together strongly.

These main kinds of inter-atomic attraction or cohesive force are not often found in 'pure' form. Most substances cohere because of a mixture of contributions from several of these forces. However, many theoretical physicists would prefer not to speak in this way of inter-atomic forces as pure or mixed, but would regard them all as manifestations of particular ways in which electrons can distribute themselves over one, two or more atoms. As we shall see in the next chapter, the theoretical problem is not simple, because the relatively simple laws governing ordinary electric and magnetic behaviour do not apply on the atomic scale.

Chemical Behaviour

We have seen that chemical behaviour depends essentially on the precise arrangement of electrons and that there are certain number relationships in these arrangements. At one stroke this discovery gave reality to the conjectures of generations of chemists. For it had been known for many years that chemical behaviour had a periodic distribution among the elements. In the Periodic Table, groups of elements of similar chemical behaviour are picked out by tabulating the elements in a number of columns (Fig. 1).

Thus the tendencies of different atoms towards particular types of chemical behaviour are in principle understood, although the degree of complexity of molecules makes theoretical prediction difficult. Some of the complex molecules now known, particularly those encountered in biology and in the new field of chemical industry associated with 'plastics', may contain thousands of atoms.

We next wish to understand why chemical reactions are so common, that is, why molecules can sometimes interchange atoms and form new substances, and why molecules already formed can sometimes dissociate again into separate atoms. The answer is to be found by considering in rather more detail the significance of the energy relationships involved and the role of temperature. Without defining temperature rigorously, we can assert that the most important consequence of change of temperature is that the amount of energy associated with an atom, molecule, crystal, droplet or other system changes in the same sense. High temperature means high energy and low temperature low energy, although in any collection of atoms or molecules the different particles do not all have the same energy; there is a distribution of energies over a range (see, for example, Fig. 18). In a gas, for example, the average kinetic energy and the average velocity will be higher at high temperatures. The tendency of a pair of atoms to approach to the distance of lowest energy of interaction will clearly be affected by what the temperature actually is. If we consider as an example a mixture of sodium and fluorine at such a high temperature that both elements are gaseous, we can safely assert that there will be molecules of sodium fluoride present *and* free sodium and fluorine. Then at higher temperatures the proportion of free sodium and fluorine will increase. Also, the proportion of the fluorine which is in the form of single atoms rather than in molecules (F_2) will increase at high temperatures. The precise composition of such a mixture could in fact be worked out if we knew all the quantities of energy actually involved.

In more complex chemical reactions in which atoms are interchanged between molecules there are more quantities of energy to be considered. There are the energies which have to be supplied before the separate molecules can be dissociated (corresponding in a simple case to the depth of the potential trough in Fig. 19) and also the energies gained by formation of the new types of molecule. Even in a gaseous mixture there will be ample opportunity for interchanges during collisions between molecules. The course of a given chemical reaction will depend on all these quantities of energy and upon the temperature, and also on the concentrations in which the various substances are present. In principle the problem of deciding which reactions will actually occur is well understood by chemists. In practice it is of course complex, but fortunately there are methods by which many of the relevant quantities of energy can be measured directly. The question is of such great importance in industries based on chemistry that the effort to understand them, using this semi-theoretical approach, has proved very worth while.

The Solid State

As has already been mentioned, inter-atomic attraction can cause atoms to condense into large lumps of matter as well as into molecules. The body of knowledge which has grown up about the actual arrangement of atoms in solids is now so immense that it is surprising to recall that almost nothing was known about it before the First World War. Most of the information has been derived by the use of X-rays, whose wavelength is of the same order as the normal diameter of an atom. We have not the space to discuss the methods used and must merely give the results; we should remark, however, that they depend on observing regularities in the spacing of atoms rather than on making observations of the positions of single atoms. The photograph, taken by X-rays, which the X-ray structure analyst obtains and studies, contains a number of spots, curves or lines, each referring to a particular regularity in spacing, and he has to interpret them by complex methods. These X-ray pictures are of course quite different from those used in medical diagnosis, which depend merely on the differing transparency of different substances to X-rays. A familiar observation which points to the basis of these methods is the pattern which can be seen when a street lamp is viewed through the fine and regular cotton network of an umbrella cover.

Starting with the simplest structures, we now describe some of the ways in which solids are built up. The structures of the solidified inert 'gases' (neon, argon, etc.) and of metals such as copper (in which there is one extra electron above a completed shell) turn out to be very simple. For the force of attraction between these atoms acts equally in all directions; it has no special directional tendency. The structure is then determined only by the requirement that each atom should be as close to its neighbours as possible. In fact, two arrangements which give the closest possible packing of spheres – that is, which pack more spheres into a given volume than any other arrangement – are adopted by these solids. Some features of these 'close-packed' structures are illustrated in Fig. 20, and can be easily understood with the help of a few marbles or ball-bearings. If we start by laying marbles on a flat surface as in (a), so that each marble touches six neighbours and is at the centre of the regular hexagon formed by the lines joining their centres, we have correctly imitated one of the basic features of these structures – the arrangement within certain crystal 'planes'. (It is advisable to deposit this layer in a 'tray' or other container which will keep the marbles in close contact with each other.) We proceed to build up the model of the crystal in three dimensions by superimposing one such plane layer on another. If we drop a marble somewhere upon the first layer, it

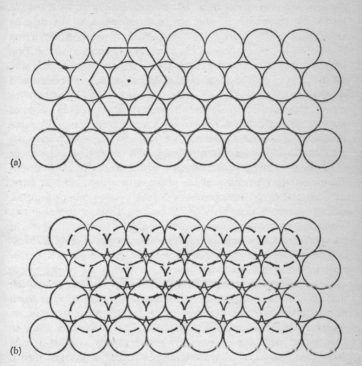

(a)

(b)

Fig. 20. (a) *and* (b) *Models illustrating the building of close-packed structures*

will naturally take up a position in one of the hollows formed between the marbles of this layer. The positions to be adopted by all other marbles in the second layer are then determined, as shown by broken lines in (b). Now in placing the third layer upon the second we find there are two possibilities. We start by placing

111

a marble in one of the hollows formed between the marbles of the second layer. With respect to the second layer there is nothing to choose between any of the hollows, but if we look at the possible positions with the first layer still in mind it is found that the new marble can either be placed immediately above a marble in the first layer, or in another position which is not directly above any marble in the two layers already deposited. If the first position is adopted, the new layer is really equivalent to the first layer and we can describe the method of putting down the successive layers by the symbol 121212. If the second position is adopted the lateral position of the third layer can be identified as 3, and the method of deposition described as 123123. This of course implies that the fourth layer is put with its centres immediately above those of the first layer, and it follows from what has been said that it could obviously have its centres above those of the second layer. In fact, except for mistakes (which do sometimes occur!) the natural arrangements are 121212 ... and 123123 ... For the arrangement in a real crystal – made of atoms, not of marbles – does not depend on putting down layers in this somewhat casual way, but is determined by the actual amounts of energy involved in the forces between atoms, and these forces extend to atoms at a distance of several layer thicknesses.

If the reader has carried out the experiment, he may note several features of interest about his pile of marbles which are very difficult to illustrate in two-dimensional figures. First, each marble touches twelve neighbours; this is the maximum number of spheres which another sphere of the same size can be made to touch. Secondly, there are many other layers now discernible in the model, inclined at various angles to the layers put down. With care, he might be able to make the upper parts of the model slide horizontally relative to the lower layers; and he can perhaps imagine how in a real crystal sliding might be able to occur in several directions. This is a rough imitation of what happens when a metal wire is severely stretched!

It is perhaps surprising that typical and very common metallic structures can be imitated so easily. There is one other which is more difficult to illustrate, in which each atom has only eight close neighbours, situated relative to itself as the corners of a cube are to the centre, but six other neighbours not too far away, being at the centres of the six adjacent cubes which share corners with its own cube. This arrangement is known as the 'body-centred cubic' structure. In this case the advantage (in energy) of having the further six neighbours offsets the disadvantage of having only eight very close neighbours. Which of the three arrangements is adopted by a particular metal cannot yet be predicted in detail. The differences in energy between them are very small compared with the total energy of interaction and such fine details are at present beyond the reach of theoretical prediction.

Crystals

The word 'crystal' has already been used here and it may be that it has previously signified to the reader something like a diamond, or quartz crystal, with smooth faces and sharp edges – quite unlike his pile of marbles. But the real significance of the term 'crystal' is that it implies an internal regularity of structure such as illustrated in these examples. The shape of the model as a whole is not very relevant, being superficial in more senses than one. How then is our model related to the actual arrangement in, say, a lump of metal? The answer is that a piece of metal is normally made up of a large number of crystals oriented in all directions at random, but firmly stuck to each other. If we choose a line perpendicular to the tray upon which the marbles are laid as defining the characteristic direction or orientation for the particular crystal model, then there will be such lines pointing in all directions in a correspondingly large-scale model of a lump of metal.

This polycrystalline structure, as it is called, is characteristic of metals. It arises in the first place usually because the lump of metal is formed by the freezing of molten metal. This process starts at a very large number of points within the liquid and the crystals, beginning with varying orientations, grow until they meet each other. It is possible by careful preparation of a metallic surface to see the individual grains with the help of a microscope. A typical photograph is shown in Plate X in which the separate grains, each having an irregular shape, can be clearly seen. The average size of a single grain might be about 10^{-3} millimetres

– about 10⁴ atomic diameters – in length or breadth. The fact that
they are then able to stick firmly together, in spite of the dis-
continuities in arrangement which must exist at the crystal
boundaries, is mainly due to the generally non-directional
character of the inter-atomic attraction. This is also the reason
why alloys, or solutions of different metals in each other, can exist
over such a wide range of compositions. As would be expected,
the inert 'gases' can also be solidified into polycrystalline
lumps, because their inter-atomic attractions are also non-directional.

Although one cannot normally say anything about the crystalline structure of a lump of metal by looking at it with the naked
eye, it is possible by special techniques to make neighbouring
crystals grow at the expense of each other until, ultimately, the
whole volume is occupied by a single crystal which has, so to
speak, 'eaten up' all the others. The properties of these large
metallic single crystals are extremely interesting. For example,
if a bar consisting of one crystal is severely stretched, markings
appear on the surface (Fig. 21) which suggest that during the

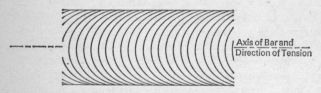

Fig. 21. Appearance of slip-lines on a single crystal

extension of the bar sliding has occurred on a number of parallel
planes in the crystal. In the early studies on such specimens it was
puzzling to the experimenters to find that the bars were sometimes brittle and sometimes quite tough. But when the specimens
were studied by X-ray methods it was realized that the crystal
orientations differed from one bar to another, and that the
behaviour of a particular bar must depend on whether a plane
upon which easy sliding could occur was inclined suitably with
respect to the direction of the force applied. If it could not slide

easily, then it might break completely along some other plane. Of course, if the large crystal is obtained by a method which depends upon one crystal eating up all the others, it is hardly possible to decide at the start which crystal will win.

Turning from the simplest structures we proceed to describe the main features of the ionic crystal. The most characteristic and common example is sodium chloride, or common salt. This is made up of positively charged sodium atoms, or ions, and negatively charged chlorine ions. The determining factor here is that ions of opposite charge attract each other but ions of like charge repel each other. Each ion, therefore, tries to surround itself by as many as possible of its opposite numbers and to keep away from ions of its own kind. The result is shown in Fig. 22, in which the

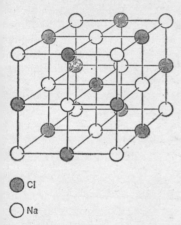

Cl

Na

Fig. 22. The structure of sodium chloride

ions have been represented by circles whose diameter is small in comparison with their distance apart. This is not done in order to suggest that the atoms in ionic crystals are more openly spaced than in metals, but only in order that the arrangement in three dimensions can be more easily understood. In reality, the ions must still be imagined as quite closely packed together. In such a

115

crystal we cannot anywhere see a 'molecule' of sodium chloride, as would be represented by NaCl, where Na stands for a sodium atom and Cl for a chlorine atom. No single sodium atom belongs to a single chlorine atom; they all belong to each other. It might perhaps be said that the whole crystal is now a molecule, if one likes to think in chemical terms.

The 'rock-salt' structure, as this is called, is probably the simplest ionic crystal structure. In general, however, the structures of ionic crystals cannot be as simple as those of metallic crystals, for there are always at least two kinds of atom to be accommodated and they may be of differing size. Indeed, quite complex structures are very soon found, particularly if there are more than two. The structure of calcium fluoride (CaF_2) is illustrated in Fig. 23. Calcium normally has two electrons over the comple-

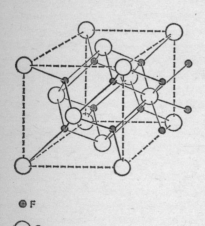

⊙ F

◯ Ca

Fig. 23. The structure of calcium fluoride

ment of the largest filled shell and thus has a tendency to form doubly charged ions. Since the ions of fluorine can be only singly charged there must be twice as many fluorine ions as calcium ions in order to preserve the net balance of charge. The degree of complexity occasioned by this rather simple requirement is rather striking; the structure is already more complicated than one can easily apprehend.

As might be expected from the selective nature of the inter-atomic attraction, ionic crystals do not easily stick to each other and polycrystalline lumps of such substances do not occur. Large crystals can however be grown from solutions, or sometimes from the molten substance, and they usually have the characteristic 'crystal' appearance in which plane faces and sharp edges can be seen. As has already been seen from the example of the pile of marbles, there are very many identifiable 'planes' in a crystal. Which planes come to the fore and are seen in the finished crystal of regular shape is a matter which may depend upon differences between the rates of growth in different directions. A curious fact is that the shape or 'habit' of natural crystals found in the ground may differ according to the locality. This is because minute traces of other substances, which may or may not be present in the locality, can sometimes become attracted preferentially towards particular faces of a growing crystal and inhibit growth in a particular direction.

As soon as we consider elements in which the 'electron-sharing' type of force is important, even more complex structures are found. For there is a directional tendency in this type of force. The atom behaves as if it had a particular number of arms (or bonds, as they are usually called) reaching out in particular directions. Except for a small number of crystals such as diamond (consisting only of carbon atoms) in which the requirements of each atom as to the preferred number and position of its neighbours can be satisfied exactly, the tendency is towards more and more complex structures, whose elucidation requires all the art and skill of the X-ray crystallographer. In some crystals molecules occupy sites in the crystal framework (or lattice) just as atoms or ions do in other crystals. Structures can indeed be encountered which depend for their geometrical arrangement on all the considerations mentioned above in the course of discussion of the different kinds of inter-atomic force.

Properties of Crystals

Let us now consider briefly some properties of crystals of the types so far described. As soon as the structures of the simpler ionic crystals became known, physicists tried to work out 'theoretical' values for various measurable properties in order to compare them with the known measured values. In the case of sodium chloride, for instance, the charges and distance apart of the ions are known and it might be expected that one could work out accurately the magnitude of the inter-atomic attraction. This leads fairly directly to an estimate of the compressibility or other elastic properties of the crystal. Theoretical and experimental physicists devoted much attention to the case of sodium chloride and arrived at the following surprising conclusions: while the experimental and theoretical values of the compressibility agreed quite well, there were serious discrepancies in the estimates of the force required to cause sliding of one plane over another and of the force required to cause complete fracture – both obviously within the scope of suitable mathematical treatments. It appeared that similar discrepancies existed for all solids and that, in fact, solids could show large deformations or fracture under forces very much smaller (by factors as high as 1 000 in some cases) than expected. Single crystals in particular were exceptionally 'soft' and weak.

These anomalies are now known to be intimately connected with 'defects' or faults of many kinds which can exist in crystals. Indeed, the diagram of Fig. 21 has already given one clue. For in the specimen illustrated sliding has obviously occurred only on quite widely separated and preferentially chosen planes. The spacing between 'steps' is at least of the order of hundreds or thousands of atomic diameters. In some way these planes must offer reduced resistance to sliding.

A great deal of ingenuity has been devoted to this problem by suggesting ways in which defects such as 'missing' atoms, 'extra' atoms or missing or extra lines of atoms, as well as defects of many other kinds, can affect the mechanical properties. It is easy to demonstrate one way in which a defect can reduce the strength by cutting a notch at the edge of a strip of paper with a pair of scissors. This reduces greatly the force required to tear the strip in two. More important, however, is the fact that the motions of these defects, or 'dislocations' as they are now generally called, can cause large-scale motions of planes in the crystal relative to one another to take place under quite small stresses. A great deal of progress has been made, both in working out theoretically the properties of the various dislocations which can occur in crystals and in actually photographing them in motion. Being little larger than atomic in scale, dislocations are almost beyond the reach even of the electron microscope, but in certain cases they can be seen. For example, a *line* dislocation which runs right across the thickness of a thin metallic film can be seen if viewed almost end-on. In Plate XI some of these are shown terminating on the slip lines visible on the surface. Very beautiful moving pictures have been made showing how such lines travel about the film when it is stressed.

Naturally, there has been a great deal of speculation as to whether ideal crystals, with no dislocations, could be grown by suitable methods. They might be expected to have enormous strength and to provide the raw material for new engineering materials. However, the prospect seems unlikely. This is partly because the processes involved in crystal growth are themselves affected by dislocations; the presence of certain types of dislocation can favour growth in certain directions. There is, so to speak, an advantage for the crystal in having such dislocations. But an interesting discovery, the growth of crystal 'whiskers', may indeed lead ultimately to the development of important new materials. When crystals are grown directly from the 'vapour' under certain conditions, small and very fine whiskers are observed to form which are very strong and rigid: far more so, in relation to their size, than ordinary crystals. It is generally assumed that each whisker contains a single long dislocation of a type known, and well described, as a 'screw' dislocation, running along its length. The growth of the crystal in the form of a filament is controlled by the presence of the screw dislocation. It is strong because there are no, or few other dislocations present, and because the screw dislocation does not easily move in the crystal. Unfortunately, although crystal whiskers are probably stronger than any other form of matter, they can still be produced only on a microscopic scale, no more than a few millimetres in length.

The grain boundary is a particularly important factor since it offers resistance to sliding and to fracture because of the sudden change of orientation of planes in the crystal, and the introduction of more and more boundaries by reducing the grain-size increases both the resistance to stretching and the ultimate strength of metals. This is why such processes as rolling and forging are carried out in the manufacture of metals. The severe deformations which these cause break down the crystals into smaller and smaller grains, ultimately reaching a limit because of the inherent tendency of the metal to 'heal' up to a certain grain-size. The metallographic examination of surfaces in order to obtain pictures such as that of Plate X is very important to the metallurgist since it enables him to find whether the grains in a metal specimen are of uniform size and whether the size is that desired.

Graphite is an interesting substance in its mechanical properties. It has strong cohesion within certain atomic planes but weak cohesion between the planes. Sliding of these planes relative to each other gives rise to its 'slippery' characteristics.

Disordered Structures

Although crystallinity, that is, the existence of a regular or ordered internal structure extending at least over thousands of normal inter-atomic distances, has usually been considered the characteristic feature of the solid state, it is surprising to find how few of the common materials fit easily into this category. What are the structures of paper, wool, plastics, rubber, glass, coal? And what are typical structures of biological systems? X-ray methods have been increasingly used in recent years upon the study of these materials, and it is known that they all contain, to a greater or less extent, amorphous or disordered matter, lacking the characteristic ordered arrangement of crystals. The key to many of these structures lies in the ability of carbon atoms to take part in extremely long molecules having a chain character.

Indeed, the study of the compounds of carbon, which is in itself a major fraction of chemistry, has been given the name 'organic chemistry' because of the importance of this element in biological organisms. In the case of diamond, carbon atoms behave as if they have four bonds reaching out towards other atoms. This can lead to a chain of the kind suggested by

$$-\overset{|}{\underset{|}{C}}-\overset{|}{\underset{|}{C}}-\overset{|}{\underset{|}{C}}-\overset{|}{\underset{|}{C}}-$$

in which each carbon atom takes part while still having two bonds to spare. Other atoms or groups of atoms can attach themselves to these to form substances of very many different kinds. A very common example is polythene, now used for electrical insulation and for kitchen ware, in which two hydrogen atoms are attached to each carbon atom. Other well-known substances depending upon such chains or upon rather more complex chain structures are cellulose, nylon and rubber.

Chain molecules can be many thousands of atoms in length and it is upon this fact that very much of the behaviour of these substances depends. These molecules are so long and complex that often they simply cannot fit themselves into any ordered arrangement. A lump of such a substance is just a bundle of chain molecules, as illustrated schematically in two dimensions in Fig. 24 (a). The bundle stays together partly because the molecules are to some extent knotted, or tangled, and partly because they are in any case 'sticky' through the inter-atomic attraction of secondary types, acting at a very large number of points. Often very high strength is found in these substances (in cotton and nylon, for instance); they sometimes appear not to suffer from weakening effects as do crystals and to exhibit values of strength which correspond to the real force required to pull chain molecules apart.

High polymers, or plastics, are of this general type or of that shown in Fig. 24 (b). Here cross links are established between the atomic or molecular groups of different chains and the structure becomes three-dimensionally braced. Such materials (of which Bakelite is an example) are on the whole more likely to be brittle than those of class (a). In some substances there is an intermediate

structure in which a greater or less amount of local regularity i
found, as illustrated in Fig. 24 (c). Over small regions the chain
lie approximately parallel but other regions are again oriente
at random. An interesting case is that of rubber, in which th
development of such ordered regions (or 'micelles') can b
clearly seen in the X-ray picture when the rubber is stretched . I
the rubber is allowed to recover its normal shape they disappear

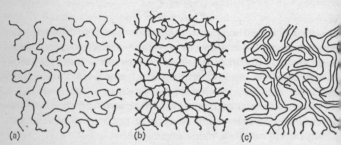

Fig. 24. Chain molecules: (a) *without cross links;* (b) *with cross links;*
(c) *with ordered regions*

The science of *materials* has developed greatly in recent years
and important materials of new types have resulted. Indeed
industry now relies much more upon synthetic materials that
upon materials which occur naturally. Many of the syntheti
materials are 'composite'; in this, of course, they imitate nature
We shall mention just two, to illustrate the way in which pro
perties can be 'blended' to lead to the desired result of produc
ing very strong materials.

'Fibreglass', a very well-known, strong and tough material
is a composite made up of fine glass filaments embedded in a
resinous plastic matrix. Glass fibres are very strong, for reason
somewhat similar to those which account for the high strengtl
of cotton and nylon, and of crystal whiskers. They are also ver
highly resistant to corrosion. The composite material, in whicl
the plastic, so to speak, holds the glass fibres together, and
allows them to use their strength, is an admirable material o
great versatility.

The second makes use of fine *carbon* filaments, also embeddec in a plastic matrix. Carbon fibres are stiffer than glass fibres, and are far less sensitive to rough handling during fabrication. In fact carbon fibres are second only to crystal whiskers in strength and stiffness, but unlike them, can be produced without limit of length. The method of production is surprising, because its starting-point is the carbon chain which we have just described as the basis of so many substances.

Plastic materials such as nylon can, of course, be drawn into very fine fibres, in which the carbon chains are orientated preferentially in the direction of the fibre. For the production of carbon fibres, plastic fibres are stretched on a frame, and heated in a furnace. Eventually everything is driven off but the carbon, which settles into a new structure. Each fibre now consists of a bundle of filaments in which the structure is closely related to that of graphite in its 'strong' directions, together with a certain amount of amorphous carbon.

After cooling, the carbon fibres are bonded into the plastic matrix. The material which results is as stiff as steel but five times lighter, and has many other desirable properties, such as a total resistance to corrosion by water. It has begun to revolutionize the engineering industry, and has already been used in some of the most demanding situations: for aircraft and spacecraft frames, propellers and bearings.

Liquids and Gases

When a solid is heated to high temperatures the average energy of the atoms or molecules increases progressively. At first this energy is associated with increased vibration about the normal sites, but at sufficiently high temperatures, the atoms or molecules break away entirely from each other and their energy is then mainly the kinetic energy of motion. This is the gaseous state of aggregation of matter. The temperature to which the solid has to be heated, or conversely, the temperature to which a gas has to be cooled before it will solidify, depends upon the strength of the inter-atomic attraction. The inert gases do not solidify until quite low temperatures have been reached, while some metals remain solid until very high temperatures.

To a first approximation there is no energy of interaction in a gas unless it is highly compressed, and most of the properties of gases can be deduced theoretically by treating the atoms or molecules as hard particles flying about constantly and making collisions with each other and with the walls of the containing vessel. If there is no vessel or other containing agent (such as the gravitational force of the earth upon its atmosphere) the gas expands indefinitely.

Most solids first melt into the liquid state before becoming gaseous. This is not true under all circumstances, as we shall see shortly, and it is convenient for the moment to consider the case of the solid–gas transformation such as is ordinarily observed with, say, solid carbon dioxide (a substance often used in the storage of ice-cream). It is hardly sufficient to regard the above statements as explaining the difference between solid and gas. It is also necessary to understand why a quantity of a given solid and a quantity of its gas can be in equilibrium with each other at a certain pressure and temperature – the pressure depending upon the temperature – even though they do not have the same energy per molecule, and energy in the form of heat has to be supplied to convert more of the solid into gas. This is one of the most puzzling facts of elementary physics. That is, we might have expected that the change from solid to gas would consist essentially of an exchange between the potential energy of the solid and the kinetic energy of the gas. The problem is solved in the science of thermodynamics, which concerns the relationships between heat and other forms of energy. We find that it is not simply the energy which determines the conditions of equilibrium in such cases. We have, of course, to take the energy into account, as we did in discussing the inter-atomic distance at which atoms would find themselves in equilibrium with each other. But we must also take into account the tendency, in nature, for systems to become disordered if given the opportunity. It is possible to set up a measure of the disorder of a system using a quantity called the entropy and usually given the symbol S. When a solid and its gas are in equilibrium with each other at a given pressure and temperature, it is found that the so-called 'free energy' must be the same for both (for, say, unit mass of each). The free energy is given by the expression $H - TS$, where T is the absolute temperature, and H is the quantity of heat which would have to be put into a given system to raise it from the absolute zero to the given temperature at the given pressure.

We now see, by inspection of this expression, which is a difference between two terms, why a gas with high energy and high disorder (or entropy) can be in equilibrium with a solid, having lower energy and lower disorder. Further, we see that there must be a general tendency of all matter, as the temperature is increased, to become more disordered as well as more energetic.

Liquids are intermediate between solids and gases in their disorder. It will be useful to study the typical phase-equilibrium diagram of ordinary substances, illustrated in Fig. 25, which defines more precisely than we have so far done under what conditions the solid, liquid or gaseous state is found in a particular case. The lines of this diagram define the relationships between the particular pressures and temperatures at which pairs of phases can be in equilibrium. Thus the line labelled solid–vapour tells us the pressure at which solid and vapour are in equilibrium at any temperature. The areas between the lines then represent a range of pressures and temperatures at which only one phase exists. (For simplicity we ignore certain distinctions made between the terms 'vapour' and 'gas'.) Familiar ideas about change of phase – that heating causes solid to turn into liquid and liquid into gas – can be seen to be valid only within a certain range of pressures. For carbon dioxide, the line corresponding to atmospheric pressure would be below the point at which all three lines meet (the triple point) and the solid then turns directly into gas.

Another point to note in this figure is the critical point. It will be seen that above a certain pressure the distinction between liquid and gas disappears and the only change of phase which takes place is that from solid into gas – or it is better to use the term 'fluid', since the material is in a state which cannot be identified definitely as liquid or gas. Of course, when both liquid and gas are present it is easy to identify which is which since the liquid is denser and falls to the bottom. When only one phase exists, however, one cannot always be certain what to call it since a highly compressed gas has the same kind of molecular arrangement as a liquid. Such a highly compressed gas is of course very different from the 'ideal' gas, in which no inter-atomic attraction exists.

The physicist is not interested only in the behaviour of matter at atmospheric pressure. He can create pressures in the laboratory as high as 10 000 or even 100 000 atmospheres in order to study the more general behaviour of matter. He then finds that the solid–fluid line of Fig. 25 continues upward indefinitely to

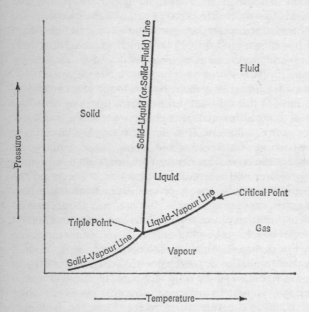

Fig. 25. Phase-equilibrium diagram of a simple substance

higher and higher pressures. This might be thought to reduce somewhat the importance of the true liquid state, which now can be seen to have a separate existence over only a relatively small range of pressures and temperatures; it is necessary, however, to try to understand how liquids, or highly compressed gases, differ from solids and from rarefied gases. The difference, as has already been implied, is in the degree of order. X-ray studies show that no regularity can be detected over a range greater than one or two atomic diameters; there is only the short-range order consequent upon the fact that the atoms or molecules have more or less precisely defined sizes and shapes. The complete explanation of the form of Fig. 25 would be very complex; in particular, there is not yet real agreement as to why there exists a critical point above which the liquid and gaseous states merge but apparently no corresponding critical point above which solid and fluid merge. This is bound up with the fact that there is a large gap between the longest range over which regularity (or order) can ever be detected in a liquid and the smallest range over which it is found in a solid – the grain-size of the finest polycrystalline aggregate.

The most characteristic property of liquids is their viscous flow. In a liquid there is a resistance to flow, or viscosity, if the flow entails relative motion between adjacent layers. In a tube, for instance, there is resistance because the layer touching the wall is at rest and layers nearer the centre are in motion. It is found that the force required to maintain a given speed of relative motion is proportional to the speed. This is perhaps the fundamental problem which must be explained by any theory of the liquid state. In a qualitative way it is easy to understand how it arises if we think of the flow in a liquid as being due to the 'rolling' of molecules or atoms around each other under the action of the applied force. It follows also if we think of a liquid as being a polycrystalline solid of very small grain-size (very much smaller than the smallest grain-size ever found in a real polycrystalline specimen) and recall that even in such a solid there is a tendency to 'heal' where grains are severely deformed or broken up by deformations.

Glass is especially interesting because it is structurally almost exactly the same as certain types of liquid, having no long-range order. This accounts for many of its properties, such as its transparency and its capacity to soften progressively when heated. Of course in its ordinary properties it behaves as a brittle solid rather than an ordinary liquid. However, when its behaviour is examined in detail it is found that many of its properties are more typical in kind, though not in magnitude, of liquids rather than of solids. In particular, the smallest force will cause glass to 'flow' if sufficient time is allowed (even though years might be required before a measurable amount of flow occurs) and this is typically liquid behaviour. The anomalous status of glass is then explained by saying that the difference between a glass and an ordinary liquid is that its viscosity is much greater – by a factor of perhaps 10^{22} at ordinary temperatures.

CHAPTER 6

From Classical to Modern Physics

Classical Physics

The structure of physics at the end of the nineteenth century, which is more or less coincident with that part of the subject now usually known as 'classical physics', presented an impressive – and apparently impregnable – façade to the scientist of the day. Indeed, it was often asserted that physics was coming to an end! The overthrow of this complacent idea has been so complete that physicists will probably refrain in future from repeating the error.

Before proceeding to extra-terrestrial questions we trace in this chapter the main direction taken by physics as a whole since the end of its classical period. Not only will this dispel any suggestion that the only developments of note have been in atomic or nuclear physics, but it should also help the reader in his understanding of what is said in the other chapters of this book and guide him towards a unified view of natural phenomena.

Classical physics had indeed achieved a remarkable synthesis. The laws of mechanics, combined with Newton's law of gravitation, accounted for the motions of the planets and stars with almost perfect completeness. The laws of electromagnetism were adequate to account for all ordinary electric and magnetic phenomena and had predicted the existence of electromagnetic waves. By 1887 it had been shown that such waves existed, and later it was shown that they travelled with the same velocity as that of ordinary light. Indeed, by the turn of the century a great deal of the complete 'spectrum' of electromagnetic waves, illustrated in Fig. 26, had been investigated.

In spite of their almost complete ignorance of the properties of individual atoms, workers in the branch of physics known as statistical mechanics had been able to make use of the atomic hypothesis in order to account for the most important properties of gases (in the kinetic theory of gases), and to a large extent for thermodynamic laws. This was achieved by combining the principles of mechanics with certain mathematical methods similar to those used by statisticians or insurance companies who have to deal with large populations. There was certainly good reason to feel that most of the natural phenomena known to physics had been neatly identified, labelled and correctly arranged in relation to each other.

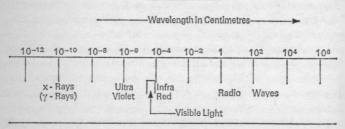

Fig. 26. Range of electromagnetic radiation

Nevertheless, there was no real synthesis between mechanics and the study of matter. Physicists were also somewhat self-conscious about the lack of synthesis between mechanics and electromagnetism. For while all other known types of wave, such as transverse waves on stretched strings or longitudinal sound-waves in air, required a *medium* for their transmission, electromagnetic waves could travel through a vacuum. The difficulty had been so acutely felt that the term 'ether' had been coined for a hypothetical medium, of which no properties could be discovered besides its ability to transmit electromagnetic waves.

Let us now see how the cracks in the façade were prised open and how, with the simultaneous growth of atomic and nuclear knowledge, physics has been remoulded into a new and stronger unity.

Relativity

The theoretical structure developed by Einstein and usually known by the term 'relativity' is very little understood by the public, although highly popular. In a sense it exemplifies the layman's concept of science as being somewhat absurd, since it can apparently lead to such statements as the assertion that the mass of a body can vary with its velocity, or that there is no such thing as simultaneity. In the hands of some writers on popular science the apparently fantastic aspects of Einstein's discoveries have been purposely stressed, whereas it should be emphasized that the ideas of relativity constitute no more than a sober recognition of certain observed facts of nature.

The problem which Einstein set out to solve concerned the 'ether', which was supposed to transmit light and other electromagnetic waves. What was the state of motion, if any, of the ether? It seemed unlikely in the extreme that the ether could always be moving exactly with the earth and so it must have some velocity, relative to the earth, which would be expected to differ according to the position of the earth in its orbit around the sun. Now according to classical ideas, the transmission of any kind of wave-motion occurs with a characteristic velocity *relative to the medium*; to any observer moving relative to the medium (such as an observer on a moving ship studying the velocities of waves on water) the velocity would appear to be different. So the velocity of light would be expected to differ in different directions upon the earth, according to the direction and velocity of the 'ether wind' at the time. (We are only concerned here with the velocity of light in vacuum, not with the changes in the velocity which occur when light passes through a transparent material, and which give rise to the phenomenon of refraction.) This would have many consequences which could be tested by observation. In 1887 the American scientists Michelson and Morley, in one of the most famous experiments ever performed, made a direct test of the proposition that the velocity of light might vary according to the direction of transmission.

They used a most sensitive and delicate optical instrument developed for the purpose, in which a direct comparison could be made of the velocities of light in two perpendicular directions. This could have detected easily a difference equal in magnitude to the velocity of the earth in its orbit (about one ten-thousandth of the velocity of light). The result was entirely negative. Indeed, without now discussing all the other possible ways in which an ether wind might have been detected, it can be asserted that no generally accepted observations have ever been made which indicate that it exists.

It thus appears that if we have, say, a beam of light passing across two observation stations, its velocity will appear to be the same when measured at either station, whether or not the stations are moving relative to each other. This is the central dilemma. Einstein's proposal was that we accept this assertion as a starting-point and examine its implications. It very soon becomes obvious that these must extend far beyond the original proposition. For all kinds of idealized experiments can be devised by which the assertion might be tested. These may involve mechanical or electrical devices, and we know that in every experiment something will prevent the observers at the two stations from finding different values in their measurements of the velocity of the beam of light. One might imagine, for example, that motion alters the length of a rule or the timekeeping of a clock, so that the two observers moving relative to each other would be working with apparently dissimilar apparatus. This apparently rather fanciful suggestion points to the way in which the theory was actually developed.

Einstein then postulated that not only the velocity of light but *all* the laws of physics would be the same in all laboratories, whatever their motion. In particular, if two laboratories were moving relative to each other, observers in the two laboratories would find the laws of physics to be identical. We use the term 'laboratory' to describe a station at which there would be rules, clocks, electrical instruments or other apparatus, which could have been standardized against each other. In fact, Einstein at first considered this principle to hold only for uniform relative motions – in the special theory of relativity of 1905. Later he extended it to cover motions which might include acceleration – in the general theory of relativity.

Now although it was postulated that physical laws would be the same in every laboratory, this did not mean that the physical laws already discovered were to be considered as of such perfect generality; it has already been seen in discussing the dilemma of the motion of the ether that this cannot be so. On the contrary, it appeared that many of the physical laws already in use had to be slightly modified, and only in their modified form did they obey Einstein's postulate. The procedure by which they were modified consisted of devising imaginary (or 'ideal') experiments – often of the kind which might earlier have been thought suitable for the purpose of looking for an ether wind – and making certain mathematical adjustments in order to remove the discrepancy suggested by the older laws, so as to bring them into conformity with the new postulate.

The kind of adjustment made can be illustrated by stating that mathematical expressions of the following form appear frequently in the revised formulation of mechanics:

$$\sqrt{1 - v^2/c^2}.$$

Here v is a velocity – perhaps of one observer relative to another – and c is the velocity of light. As we have already seen, v/c is only about $1/10\,000$ even when v is the velocity of the earth in its orbital motion. The above-mentioned expression then has the value 0.999999995, which is only minutely different from unity. In ordinary laboratory experiments involving much smaller velocities the departure from unity would be even smaller. This explains why the laws of Newtonian mechanics had stood for so long without contradiction. For velocities small compared with the velocity of light the old laws remain valid, and indeed, all the new laws become indistinguishable from the old laws if only small velocities are considered. But for large velocities, such as the velocities of galaxies relative to each other, or those now attainable in particle accelerators, they lead to quite different conclusions.

It should not be imagined that it was an easy matter to set up the new laws. First, they had to represent correctly all relevant phenomena as known at the time. Furthermore, the greater complexity of the new mathematical structure meant that some of the basic concepts of mechanics had to be reconsidered. For example, it was necessary to reconsider the definitions of force, mass and acceleration and at this stage a certain freedom of choice was possible. In making a choice Einstein always favoured simplicity in mathematical form, and it is the great triumph of his genius that the theory of relativity as set out by him succeeded, not only in its immediate object of representing mechanical and electromagnetic behaviour as then known, but also in making a number of forecasts which have subsequently been verified by observation.

Without attempting an exact or full account, and without reference to whether the role of the theory was to provide clarification, explanation or prediction, we now mention briefly some of the main features of relativity theory.

(i) The concept of 'absolute time' has to be given up. Newton's view was that 'absolute, true, and mathematical time, of itself, and by its own nature, flows uniformly on, without regard to anything external'. Now, however, it cannot be said that an event occurs at a given *absolute* time and accordingly absolute significance cannot be attached to simultaneity, nor can it necessarily be said for certain which of two events precedes the other if they occur at different places. It will depend on where the observer is situated relative to the locations of the two events, since the way in which he can receive the most exact information about their occurrence involves the passage of a light signal, travelling with finite velocity. If (in space) the observer is near event A, say, he might assert that it occurred before event B, whereas if he had been near event B he might have said the opposite.

(ii) In the same way, 'absolute space' is no longer a valid concept. For while one observer might say that two events occurred at the same place and at different times, another moving relative to him would say that the events occurred at different places. In a sense, space and time have clearly become linked by relativity, and this is why time is often spoken of (but without exact significance) as the 'fourth dimension'.

(iii) In the new mechanics it appears that the mass of a moving body is not the same as its mass at rest and that the gain in mass is $1/c^2$ times the kinetic energy due to the motion. Expressing it another way, the kinetic energy is c^2 times the gain in mass. Einstein's famous equation

$$E = mc^2$$

represents the idea that the ordinary 'rest' mass of a body is also associated with an amount of energy equal to its mass multiplied by the square of the velocity of light. As we have seen in discussing nuclear phenomena, this is indeed an experimental fact.

(iv) Not only have the laws of mechanics, but Newton's law of gravitation also has to be modified. This was the problem dealt with by Einstein in the general theory of relativity which, as has been said, is concerned with observers who are travelling with accelerated motion relative to each other. Accelerated motion is, of course, characteristic of bodies moving under gravitational attraction. It was in this field that Einstein's most spectacular success was achieved, for he made three predictions which have been confirmed by observation. He predicted that very slight differences from 'Newtonian' behaviour should be expected in the motions of the planets. Only Mercury is sufficiently near the sun to show a measurable difference, and the existence of a discrepancy was in fact already known to astronomers but not understood. Also, Einstein predicted correctly that light beams would be bent in a strong gravitational field (say, in passing close to the sun) and that spectral lines, which give information about the frequencies of atomic or molecular vibrations, would indicate an apparent slowing down of such processes on the sun. Some of these matters will be discussed further in later chapters

The theory of relativity forms a popular background for discussions of a metaphysical nature because of some of its apparent absurdities. But in astronomy, and in the design of particle accelerators, it is simply a necessary part of the technique, because relativistic mechanics leads to a correct account of certain natural phenomena which could not otherwise be explained. Shorn of some of its overtones of fantasy, we see that the central principle of relativity is in effect no more than a restriction upon the type of theory which can have full validity in physics. Every theory must be verifiable in every laboratory, whatever its state of motion. Perhaps its main lesson for physicists is that they should keep open minds and should be prepared to question even their most firmly established ideas in the light of new discoveries. In particular, relativity suggests that one should beware of ideas like 'absolute time' independent of the observer.

Quantum Theory

At about the same time as Einstein was carrying out his work on the behaviour of bodies of very large size or moving with very high velocities, classical physics was receiving another series of blows following some of the new discoveries in atomic physics. The first rift at this other end of the scale of natural phenomena had, however, come from an unexpected quarter when Planck proposed his quantum theory in 1900 as a result of his analysis of certain properties of thermal radiation. Subsequent developments went far beyond the scope of Planck's original theory and are distinguishable by different titles such as 'wave mechanics' and 'quantum mechanics'.

The problem studied by Planck was the distribution of energy among different wavelengths in the thermal radiation emitted by hot bodies. It had been shown earlier that there was a certain distribution of energy within any enclosure, and that this was in fact the same for all wavelengths at a given temperature, whatever the nature of the walls. It is easier to believe this assertion if it is realized that surfaces which are good emitters of radiation are also good absorbers, while bad emitters are bad absorbers, that is, good reflectors. So, although we might imagine that it takes longer to set up the required energy distribution in an enclosure with highly polished walls, this does not alter the final result. We can imagine the energy as being associated with radiation in the act of passing between the walls. Although the origin of the radiation must lie in the oscillations or vibrations of atoms (any oscillating dipole acts as a source of electromagnetic radiation), the fact that the amount of energy in the enclosure corresponding to a given wavelength did not depend on the walls suggested that the form of the distribution was in itself a fundamental fact of physics. The actual form is shown in Fig. 27 for

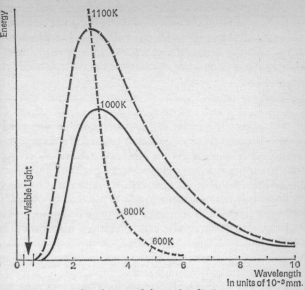

Fig. 27. Energy distribution of thermal radiation

two temperatures, 1 000 and 1 100 K. It will be seen that at the higher temperature the peak in the curve is at shorter wavelength and beginning to overlap the region of visible light. The positions of the peaks of the distribution curves for various temperatures are indicated by the fine broken line.

We have not the space to explain to the reader precisely the theoretical situation as Planck found it. The theory included a discussion of the number of ways in which stationary waves of differing wavelengths could be set up in an enclosure rather similar to the discussion which led to the kinetic theory of gases. Let us merely say that the conclusion of the theory was that the curve of energy distribution against wavelength should show no peak but rise towards infinity at very short wavelengths. Planck's new postulate was that the energy of an oscillator (such as an atom) could not vary continuously but could only take values which were multiples of the quantity hv, where v is the frequency of the oscillation and h a universal constant, which was the same

137

for all types of oscillator. This is now known as Planck's constant. Further, he assumed that the emission or absorption of radiation was accompanied by a jump between two of these 'energy levels'. If the jump was downward, from higher to lower energy, radiation would be emitted. The absorption of radiation would lead to an upward jump.

Using these postulates Planck was able to modify the existing theory so as to account exactly for the form of the curves of Fig. 27. The complete success of the new theory caused much astonishment because of the revolutionary nature of the empirical assumptions made. For clearly the jump, or transition, between two adjacent energy levels could be accompanied only by an emission or absorption of the quantity, or 'quantum', of energy hv. This suggested that radiation could be emitted in pulses, a concept which was out of harmony with the existing idea that light and other electromagnetic radiation were transmitted as continuous trains of waves. At the time this revolutionary concept had to stand, unsupported, as an *ad hoc* assumption leading to one specific and highly important result.

The next step was taken by Einstein who, a few months before the publication of his first paper on relativity, drew attention to the fact that other evidence existed for the suggestion that the exchange of energy between matter and electromagnetic radiation took place in pulses, or quanta of energy. This evidence came from the results of experiments on the photo-electric effect, which has been mentioned in Chapter 1. It appeared quite unequivocally that the exchange of energy took place in quanta of energy hv, where v is again the frequency of the incident light and h is Planck's constant. At this stage, therefore, there was the confusing situation that although light appeared to have corpuscular properties, only by thinking of it as a wave, having wavelength and frequency, could one define the energy of the corpuscle, or 'photon'. And, of course, the body of knowledge about the wave-like properties of light remained valid.

The attribution of corpuscular properties to light was not, however, entirely unexpected because it was already known that light falling on any surface exerted a pressure, the so-called 'radiation pressure', upon it. This had been observed experimentally and is also a specific theoretical consequence in thermodynamics, electromagnetism and relativity theory.

The Bohr Atom

The next important development resulted from attempts to explain the characteristic spectra of certain substances. When a discharge in, say, hydrogen gas is examined in a spectroscope the resulting spectrum is not a continuous series of colours (the rainbow sequence) but consists of a number of bright lines at well-defined wavelengths – or colours – having obviously some systematic form of variation in their spacing. Now, although many regularities had previously been detected in the spectra of various elements, it was not until J. J. Balmer found, in 1885, that the wavelengths of the hydrogen lines could be represented by a simple empirical formula that the real challenge of this problem was fairly presented to the theoretician.

The starting-point of the theory was the idea that the emission of radiation was due to atomic 'oscillators'. The growing knowledge of the constitution of atoms suggested that these oscillators might be electrons in atoms revolving around the atomic nuclei, for a negative charge made to revolve around a positive charge would indeed emit radiation with frequency equal to the frequency of revolution. But a major difficulty presented itself immediately. For the loss of energy due to the emission of radiation would cause the electron to 'spiral' towards the nucleus at ever-increasing frequency of revolution. The emission of radiations of characteristic frequency, and, indeed, the continued existence of atoms, could hardly be explained on this basis.

The immediate solution of the difficulty was due to Bohr, who in 1913 made a number of radical postulates rather similar in type and content to those of Planck's original theory. He assumed that an electron could revolve around the nucleus at a fixed distance without radiating energy, and that there was a series of 'allowed orbits' in which this was possible. Only in the transition from one orbit to another could radiation be emitted or absorbed. In the case of a jump from an orbit to one nearer the nucleus there would be emission of radiation, while absorption of radiation could cause the corresponding outward jump. The frequency ν of the radiation was given by $\varepsilon = h\nu$, where ε is here the difference between the energies of the two levels – usually known as 'quantum levels' or 'quantum states'. In the normal state of the atom the electron would occupy the orbit of lowest energy, but the absorption of radiation would raise it to a higher orbit; that is, the atom would be raised to an 'excited state'. Subsequently, radiation might be emitted by the fall from a higher orbit to a lower. At high temperatures many atoms can be excited by collisions with other atoms and then radiate energy by a fall into a lower state.

It is obvious in this case why an atom can absorb only radiations of frequency equal to those which it can emit, so that an emission spectrum is just like the 'negative' of an absorption spectrum. In fact, however, it is a universal property of all oscillators that they absorb well the frequencies which they emit well. A tuned violin string, for example, will vibrate sympathetically, or 'in resonance', if the correct note is sounded in the vicinity, and will therefore remove or absorb some of the energy in the sound wave.

By combining certain principles of classical mechanics and electrostatics with these new postulates, Niels Bohr was able to set up a mathematical theory which completely accounted for the main lines of the hydrogen spectrum. Later, other spectra were accounted for, and the question of the 'fine structure' of spectra – that is, of the 'splitting' of main spectral lines into groups of lines very close together – was dealt with successfully. It was shown, for example, that the variation of the mass of the electron with its velocity, as predicted by relativity, accounted for some of the fine structure. A good deal of supporting evidence about the existence of quantum states subsequently accumulated from other experiments.

Wave Mechanics

Once again the adoption of postulates of a radical nature had brought order out of chaos. There remained, however, a certain feeling that these postulates had been too much of an *ad hoc* nature, and insufficiently related to other physical ideas. The next developments, though no less radical, provided such a completely satisfying synthesis of all the quantum phenomena that they removed the last doubts of physicists. These were the ideas of Louis de Broglie, Schrödinger, Dirac, Heisenberg and others, which find their most complete expression in modern 'wave mechanics'. The starting-point was a suggestion by de Broglie that, just as light had been found to have certain corpuscular properties, matter might have wave-like properties. A few years later this was shown to be true by Davisson and Germer in the U.S.A. and by G. P. Thomson, the son of J. J. Thomson, in Britain. They showed that a beam of electrons striking the surface of a crystal led to a pattern much in the same way as did a beam of X-rays. De Broglie assumed further that the frequency of the 'matter wave' associated with a particle could be found by use of the equation $\varepsilon = h\nu$, which has already been mentioned, where ε is now the total energy (kinetic and potential) of the particle.

It is now seen that the reason why a simple 'particle' theory of the atom, using ordinary mechanics and electromagnetic theory, does not succeed is that the electron cannot be considered simply as a particle. It behaves in part as a wave also. Indeed, it is possible, by considering the de Broglie wavelength of a particle, to decide whether a 'classical' theory is likely to be adequate to account for its properties, or whether a theory along quantum lines will have to be developed. For a tennis ball moving with average velocity the de Broglie wavelength is minutely small in comparison with the size of the ball or of the tennis court. Classical mechanics will suffice in this case. For an electron moving over distances of the order of size of an oscilloscope or television tube the de Broglie wavelength is also much smaller than the length of path. But for an electron moving within an atom the distance is of the same order of size as the atom itself.

The full mathematical theory is far too complex and difficult to be discussed at any length here. It contains new ideas of an extraordinary character. For example, an idea of fundamental importance is that there is an ultimate uncertainty about the characteristics of any given particle which is not due to experimental or theoretical inadequacy; the particle cannot have both its position and velocity well-defined. If we imagine how we would try to determine experimentally, say, the exact position of an electron, this becomes evident. One approach would be to try to 'see' it. But this could not be done unless at least one photon had hit the electron, and it is known that this would give the electron a 'kick'. Theoretical physicists now believe generally that if something could not, even in principle, be measured or detected, it should not be regarded as significant.

The full theory, while somewhat altering the significance of Bohr's postulates, leads to a satisfying interpretation of them and gives a complete account of the hydrogen atom. The concept of orbits does not remain in quite the same form, but is still useful. The electronic 'shells' or 'sub-shells' which were mentioned in the previous chapter correspond to certain groups of electronic orbits or quantum states, and in fact most of the emission and absorption of radiation corresponds to jumps between shells or sub-shells rather than within these groups.

Unfortunately, certain mathematical features of the theory make an extension to more complicated systems more and more difficult. It is hardly possible to make rigorous calculations about complicated systems and various methods of making reasonable approximations have been developed. The matter is very important in physics because it is believed that the technique of wave mechanics offers, in principle, the solution to most of the problems remaining in the field of atomic and molecular structure, inter-atomic forces and chemical behaviour, and the properties of matter in bulk. This is one of the main battle-grounds of the theoretical physicist today.

Quantum Theory of Solids

There has been in recent years a spectacular development of a branch of physics known as 'solid-state physics', as a result of the increasing understanding of quantum theory or wave mechanics. Many of the properties of matter, such as those which have been discussed in Chapter 5, can be accounted for fairly well without drawing too heavily on quantum ideas. This is because the de Broglie wavelength of an atom in a solid is rather small (though by no means negligible) in comparison with its size. But those properties which depend on the motions of electrons in solids have to be treated as quantum phenomena. The conduction of electricity in metals consists of motions of 'loose' electrons, and this has now received a very satisfactory treatment according to wave-mechanical theory. The 'fine structure' of solids forms an increasing part of the solid-state physicist's preoccupation because it is a feature of quantum theory that the whole of a lump of solid behaves as an atom (or molecule) in one respect: there is again a series of quantum levels – though now very closely spaced – which the electrons must occupy. It is the distribution of these quantum levels which ultimately determines the properties of a solid.

The essential difference between electrical conductors and insulators, indeed, is that in conductors there are vacant quantum levels close in energy to those occupied by the electrons and to which they must jump in order to gain the energy required to travel through the solid. In insulators there are none – all the permitted 'bands' of quantum levels are full – and the electrons cannot be accelerated.

In *semi-conductors*, such as silicon and germanium, there are some vacant allowed quantum energy levels quite near the filled bands. As we shall see in the next chapter, these materials are intermediate in behaviour between metals and insulators. Their great importance lies in the fact that particular electrical properties can be secured by manipulating the distribution of quantum levels.

Very interesting results have been obtained in studies of the properties of solids at low temperatures. At first, low-temperature physics was concerned mainly with such questions as the condensation of gases such as oxygen, nitrogen, hydrogen and helium into the liquid and finally into the solid state. But it has now proved to be a most powerful approach to the study of the 'fine structure' of solids, in particular, of the distribution of quantum levels. Just as an atom at ordinary temperatures tends to be unexcited and to exist in the lowest quantum state, so the electrons in a piece of bulk matter, with much closer spacing of quantum levels, tend to fall into low quantum states at low temperatures.

The finest details of the structure of matter are associated with magnetic behaviour, in particular with the influence on the spacing of quantum states of electronic or nuclear atomic magnets. If the temperature can be lowered to about 1/100 degrees above the absolute zero (10^{-2} K) the electrons fall into low quantum states in a way which can be described as an alignment of the electronic magnets. If the temperature can be lowered to about 10^{-4} K the much weaker nuclear magnets begin to align. Conversely, if these elementary magnets can be aligned by the external application of magnetic fields, the effects themselves can be used in lowering the temperature. Very high magnetic fields must be used to reach the lowest temperature and in 1956 the first 'nuclear cooling' was achieved at Oxford, when a temperature of about ten millionths of a degree above the absolute zero was reached for a short time.

At somewhat higher temperatures one of the most surprising discoveries was that certain metals suddenly lose their resistance to the passage of electricity when cooled below characteristic temperatures.

This property of 'superconductivity' was discovered in 1911, quite near the beginning of the period being discussed in the present chapter, but hardly understood in any fundamental sense until forty years later. In discussing ordinary electrical conduction we mentioned that electrons require energy to be supplied before they can travel through a solid. In superconductors, however, a subtle and very small attractive interaction between the loose electrons, in which the crystal lattice is itself involved, permits pairs of electrons to travel through the solid without energy being supplied.

A current once set up in a superconductor will persist for ever if the metal is kept below its own particular characteristic 'transition' temperature, and provided the current is not destroyed by external means. It is easy to set up such a 'persistent current' in a *ring* of superconductor – a reminder, perhaps of the persistent motions of electrons in atoms which we met earlier in this chapter, when one remembers that in the quantum theory of solids the whole of a lump of matter is treated as though it were an atom.

The flow of electricity on a large scale without the consumption of energy is the electrical engineer's dream, and although the dream has not yet been realized in its full sense, there have been important practical applications of superconductivity, as we shall see in the next chapter.

Finally, perhaps the most surprising of low-temperature effects, the properties of helium cooled below $2 \cdot 1$ K, must be mentioned. Helium boils at $4 \cdot 2$ K but is unique in remaining liquid without solidification down to the lowest temperatures (though it can be solidified under pressure), and when it is cooled below this characteristic temperature its properties change in a remarkable way. It loses its viscosity and will 'siphon' itself into or out of vessels along a film, thick by atomic standards, which clings to every solid surface in its vicinity. The clue to much of this strange behaviour lies in the fact that the de Broglie wavelength of helium atoms at these temperatures is of the same order as their distance apart in the liquid. We might say loosely that the atoms are uncertain whether they are particles or waves! One consequence is that the volume of a given mass of liquid helium is much greater than would have been expected on 'classical' grounds.

Philosophical Implications of the Discoveries of Physics

It is tempting to say simply that there are no philosophical implications to be attached to the matters which have been discussed here, but this would be to ignore the fact that violent controversies of a philosophical character have existed in physics at least since the time of Aristotle. At any rate, the tendency among physicists is now to consider that the real content of a statement in physics lies in the statement itself, and that to speak of its philosophical implication to some extent reduces the force of what is said. Let us try to illustrate this by some examples.

One of the most important scientific controversies was that between the followers of Aristotle and Ptolemy, who said that the sun and stars travelled around the earth, which was thus the centre of the universe, and the followers of Copernicus and Galileo, who insisted that the earth travelled around the sun. These opposite views would naturally be thought to have philosophical implications. Indeed, the Copernican hypothesis was attacked on religious grounds because it was asserted that Copernicus had removed an obstacle to the idea, opposed by the Church, that the universe might be infinite. It was unthinkable that stars travelling around the earth could move with infinite velocities, as some must in an infinite universe, if all revolved around the earth.

Now what Copernicus had found was that if he assumed the earth to be a planet, like the other planets, and that all moved around the sun, a great mathematical simplification occurred. He could then account for the seasons and for many other features of the planetary motion. So he asserted that the earth moved around the sun. This idea was at the time a great scientific advance. But with our present knowledge of astronomy we know also that the sun is moving around the centre of the stellar system; when an astronomer studies distant nebulae the situation is still further complicated. So how is the earth really moving? Surely the real truth is that which is contained in the sum of all detailed assertions about relative motions. For practical purposes any suitable hypothesis may be used: a cyclist who wished to know what was lighting-up time would find the Ptolemaic system adequate.

This old controversy carries its lesson even today. Anyone brought up in the Aristotelian tradition would resist strongly any suggestion that the earth was rotating on its own axis and revolving around the sun at enormous speed. All the evidence of the senses would persuade him that this must be a fantastic invention. Eventually, after calmer reflection, he would realize that it was really a very reasonable idea because it explained all the complex planetary motions – relative to the earth – as being also quite simple revolutions around the sun. Finally, he might arrive at the more empirical view, suggested above, that the real truth was the sum of knowledge about relative motions.

Nowadays, the ideas of relativity and quantum theory appear strange. But when they have been thoroughly assimilated future students of present-day writings on their philosophical implications may wonder what all the fuss was about! If a theory leads to a simplification or greater ordering of ideas then it is useful. The physicist does not argue as to whether it is *right*. In particular, he does not worry if he cannot explain the new postulates made. We have deceived ourselves if we think that anything can be *explained* – in the mystical, absolute sense usually attached to this word. It is only because of our extreme familiarity, almost from birth, with some of the ideas of classical physics that modern physics is so puzzling. For example, the indestructibility of matter seems so obvious from all our experience that the statement that matter can be converted into energy is difficult to accept. Also, we have rather firm ideas about the meanings of such words as 'force' and 'cause', because we know how to exert forces (with our muscles, say) and we can also cause events to occur. It is disturbing to be told that force has to be redefined in order to carry through the development of relativity theory, or that force can be transmitted without the intervention of a medium, or that certain physical events appear to occur without cause – a feature of quantum theory.

It becomes less difficult to accept such changes when it is realized that the concept of force is in itself empirical and somewhat synthetic. When Newton's second law is objectively analysed we come to realize that what it does is to define force only if mass and acceleration have previously been defined. In relativity the definition of force is slightly altered and the question naturally arises as to whether such a thing as 'force' really exists. The answer is that force is what we have now decided it shall be, and we have chosen the present definition because it leads to an *all-round* simplification of the laws of nature in their mathematical formulation. That is, force, as now defined, enters simply into many other physical laws besides Newton's second law.

It is useful when considering any statement about the philosophical implications of science to apply two tests. First: is the statement framed with the same care as scientific statements usually are; do the words used convey a precise meaning, and the same meaning to everyone concerned in the discussion? At this point it is only being insisted that the same rigour in argument be employed as is required by modern philosophers – who have, as a matter of fact, been as fundamentally influenced by the objectivity and empiricism of science as by the exact reasoning of modern mathematics. But very many of the assertions which

147

have been made about the philosophical implications of science would fall to the ground when analysed in this way. Secondly: is that which is implied really contained in that which is established by the scientific experiment? This requirement seems so obvious that it is surprising to find how many philosophical assertions are made which fail to meet it. For example, it is often said that because of the element of uncertainty introduced into quantum theory by wave mechanics there is an ultimate limit to human knowledge. Even this statement can be challenged, but it will often be added that it implies that man is an imperfect creature who should not presume to probe too deeply into scientific problems. Or it may be said that, because the concept of causality seems to disappear in certain quantum phenomena, man does have the power of making his own decisions. These arguments will hardly bear examination. At the very least it should be emphasized that the first part of each assertion is about electrons or nucleons; the second part is about man.

The scientific attitude is to avoid assertions containing terms which have not been objectively defined, and to regard each assertion about scientific fact as meaning precisely what it means and as containing precisely what it contains, neither more nor less.

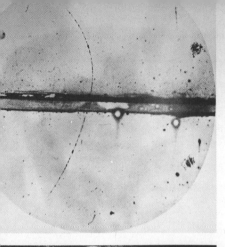

I. The discovery of the positron

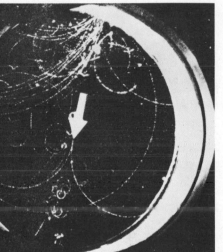

II. Pair-production

III. Transformation of a π-meson into a μ-meson into an electron

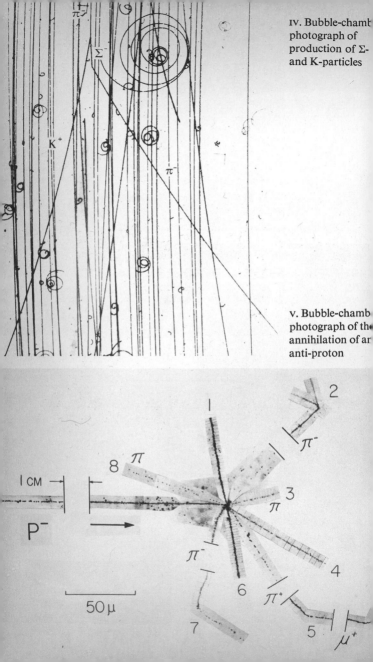

iv. Bubble-chamber photograph of production of Σ- and K-particles

v. Bubble-chamber photograph of the annihilation of an anti-proton

Tracks of α-
icles from
ium C and C′

Track of α-
icle showing
ear scattering

Scattering of
articles by helium

IX. Disintegration of nitrogen by α-particles

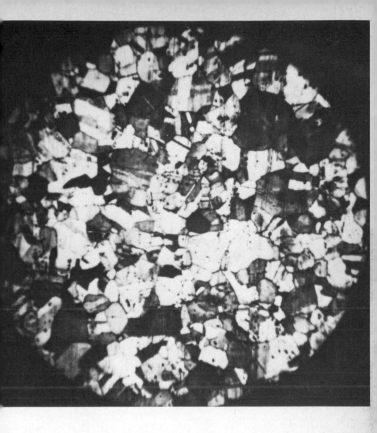

The surface structure of a metal

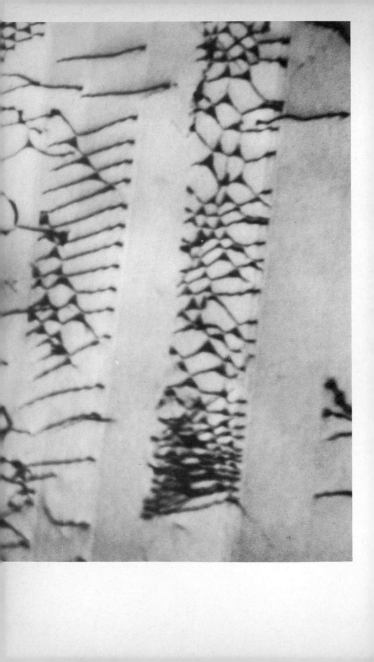

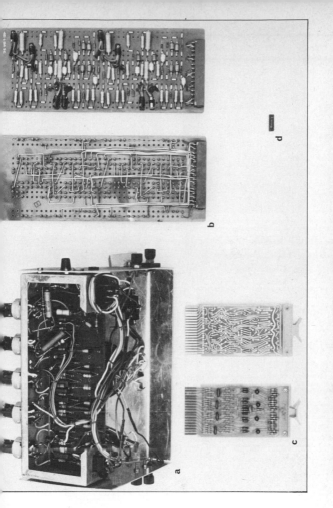

XII. A decade scaler unit: (a) using thermionic vacuum tubes; (b) using transistors with wire connections (back and front); (c) using transistors with printed circuit connections (back and front); (d) microminiaturization: a modern 'chip' which fulfils the same purpose as (a), (b) and (c)

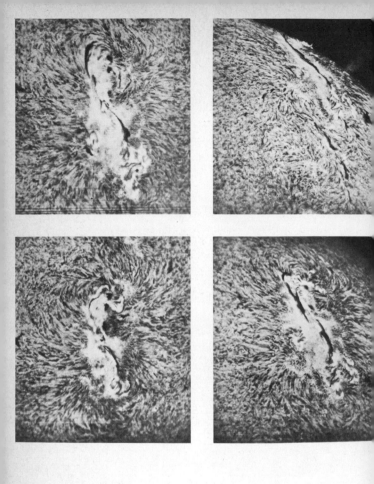

XIII. The S.W. quarter of the sun taken with the red hydrogen line (Hα) on 3, 5, 7 and 9 August 1915

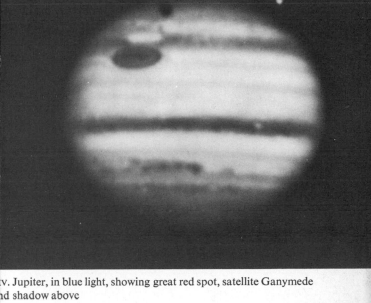

iv. Jupiter, in blue light, showing great red spot, satellite Ganymede and shadow above

v. Martian craters, taken from Mariner VI, at an altitude of 3220 km

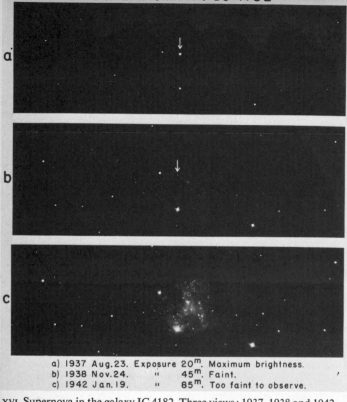

a) 1937 Aug. 23. Exposure 20ᵐ. Maximum brightness.
b) 1938 Nov. 24.   "   45ᵐ. Faint.
c) 1942 Jan. 19.   "   85ᵐ. Too faint to observe.

XVI. Supernova in the galaxy IC 4182. Three views: 1937, 1938 and 1942

XVII. The Crab nebula, in red light, Messier 1

I. Pulsar NP 0532 in Crab nebula. Two different phases

The Veil nebula in Cygnus, north part

xx. Great nebula in Orion, Messier 42

xxi. The 'Horsehead' nebula in Orion

II. Globular star-
ster, Messier 3

XIII. The Whirlpool
ebula, Messier 51

xxiv. Spiral nebula in Virgo, seen edge on, Messier 104

xxv. Relation between red-shift and distance for galaxies

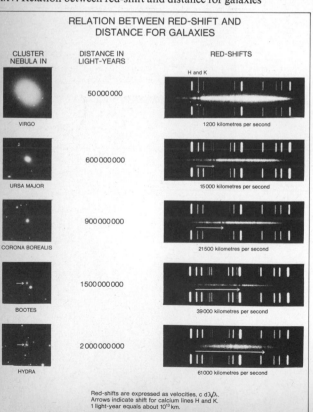

RELATION BETWEEN RED-SHIFT AND DISTANCE FOR GALAXIES

CLUSTER NEBULA IN	DISTANCE IN LIGHT-YEARS	RED-SHIFTS
VIRGO	50 000 000	1200 kilometres per second
URSA MAJOR	600 000 000	15 000 kilometres per second
CORONA BOREALIS	900 000 000	21500 kilometres per second
BOOTES	1500 000 000	39 000 kilometres per second
HYDRA	2 000 000 000	61000 kilometres per second

Red-shifts are expressed as velocities, c dλ/λ.
Arrows indicate shift for calcium lines H and K.
1 light-year equals about 10^{13} km.

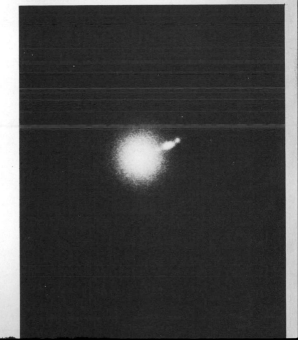

VI. Cygnus A
radio source

VII. Peculiar
elliptical galaxy,
with jet, in Virgo,
Messier 87

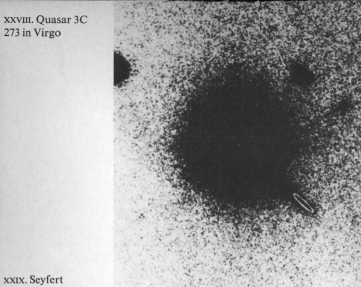

XXVIII. Quasar 3C 273 in Virgo

XXIX. Seyfert galaxy in Canes Venatici

CHAPTER 7

Applications of Electron Physics

The growth of scientific activity since the Second World War has been most dramatic in the area with which we have been dealing in the last two chapters – that of the physics of the object of medium size. This area has not been the scene of spectacular discoveries in pure physics, comparable with those of nuclear physics already discussed, or with those of astronomy, to be discussed in later chapters. The story has been, on the contrary, of spectacular *applications* of earlier discoveries. We have already mentioned two such discoveries, the properties of *semi-conductors*, and of *superconductors*, and in this chapter we shall explain some of their applications. Two other examples of the applications of physics are household words, and the reader would expect them also to be placed in context: the *computer* and the *laser*. The pure scientific basis of these two instruments is relatively unimportant on the grand map of science, and would scarcely have merited more than a few lines of this book. But the advertisement pages of newspapers are full of references to computers. How have computers, which are surely intended to 'compute', become the villains of science fiction? Why does every schoolboy know, or appear to know, about lasers?

In all four applications there is a common factor: their dependence upon the science of 'electronics', founded by J. J. Thomson in his classical experiment and developed since to an incredible level of sophistication and expertise. Of course, in semi-conductors and superconductors the 'moving parts', so to speak, are electrons, and many 'electronic' devices depend upon them. It will appear later that this is true in a sense also of the computer, but not of the laser, in which the 'moving parts' are light waves. But all depend for their successful realization on using the devices and techniques of electronics, and this is why we have chosen to discuss them under the general heading 'Applications of Electron Physics'.

Semi-Conductors

The characteristic of semi-conductors which makes them different from ordinary electrical conductors and from insulators is that there is a gap, but not too large a gap, between the bands of quantum energy levels which are fully occupied by electrons and those of higher energy – the *conduction band* – in which there are vacancies. This gives rise at once to a characteristic property, an increasing electrical conductivity at high temperatures. For electrons can be excited by thermal agitation to higher quantum states, in this case into the conduction band, and the number excited increases rapidly as the temperature is raised.

This is the basis of the *thermistor*, or 'thermally sensitive resistor', used since the early days of semi-conductors for measuring and controlling temperature. For the control of temperature such a device might consist of a thermistor connected to a supply of electricity and to suitable electronic relay circuits in such a way that when the temperature rose above that desired the increased current through the thermistor would actuate the relay and cause an electric heater to be shut off. When the temperature had fallen a little the current through the thermistor would fall and the heater would be turned on.

But the all-important developments in the field of semi-conductors are far more interesting than this. They depend, essentially, on the accurate control and manipulation of the properties of semi-conductors made possible by the introduction of *impurities*, or by 'doping'. Not only do impurities greatly increase the electrical conductivity, by providing new quantum energy levels near the conduction band, but they allow different kinds of conduction to take place, with somewhat surprising results, as we shall see.

In Chapter 5 we noted that the chemical behaviour of an

element was determined by the number of electrons in its outer shell: for silicon and germanium, typical semi-conductors, this number is four. Now if an atom with *five* such electrons, say phosphorus, can be introduced into the crystal lattice of germanium, one of these electrons is more or less redundant from the point of view of holding the lattice together, since the whole structure and symmetry are based on an arrangement in which each atom has four nearest neighbours. Only a small amount of energy is required to detach the fifth electron, which can then travel through the lattice as a negative charge – the elementary process in the passage of electricity.

If an atom with *three* outer electrons, say boron, can similarly be introduced into the germanium lattice, exactly the converse happens. Each boron atom provides, so to speak, a missing electron, or 'positive hole'. This can now travel through the lattice as though it were an elementary *positive* charge. Of course, the real mechanism in the passage of electricity is still the motion of negative charges (the electrons). For example, an electron moving to the left, say, to attach itself to a boron atom is exactly equivalent to a positive hole detaching from the boron atom and moving to the right. The behaviour of semi-conductors can however be more easily understood by distinguishing between the two types of charge carriers: electrons and positive holes.

In a typical 'impurity semi-conductor' of this kind, there might be about one atom of impurity to every million atoms of germanium, and nearly all the impurity atoms would have provided their surplus electrons (or positive holes) for electrical conduction through the lattice. Even with such a small amount of impurity the electrical conductivity might be a thousand times higher than that of pure germanium.

The two kinds of impurity semi-conductor are called 'n-type' or 'p-type' according to whether their conduction of electricity is due mainly to the passage of (negative) electrons or positive holes. Let us now consider the behaviour of two pieces of semi-conductor of opposite type in contact. In Fig. 28 (a) a piece of p-type semi-conductor on the left is in contact with a piece of n-type semi-conductor on the right, and external current-carrying wires or 'leads' are connected as shown. If a positive charge of electricity passes from left to right it will have the following main effects. In the piece of p-type, positive holes will move to the right, towards the junction. In the piece of n-type, electrons will move to the left, also towards the junction. At the junction (b) the positive holes and the electrons meeting will neutralize each other (or, in more conventional terms, the electrons moving to the left will cross the junction). At A, positive holes moving to the right provide electrons which flow out to the left, into the wire. At C, electrons flowing from the wire to the left provide the electrons which flow into the semi-conductor.

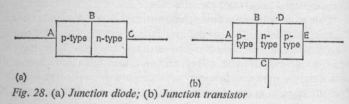

(a)

(b)

Fig. 28. (a) *Junction diode;* (b) *Junction transistor*

But if a positive charge of electricity is considered to pass from right to left, a very different situation arises. For now in both pieces of semi-conductor the charge carriers flow *away* from the junction. This region becomes 'depleted' of carriers and the current has to stop. The reader will quickly confirm that again no barriers to the flow of current exist at A or C, but only at B.

A p-n junction is therefore a *rectifier*, passing electric current in one direction only. This is the *junction diode*, the simplest device of *solid-state electronics*, the harbinger of a technological revolution. Its great advantage over its predecessor, the *thermionic valve diode*, is that heat is not required for the generation of the charge carriers.

It will be worth recalling some of the main features of thermionic valves, or 'vacuum tubes'. They consist of a number of 'electrodes' in an evacuated glass tube. One of these, the cathode, is heated, and generates a supply of electrons; the others are arranged in various configurations according to their purpose. In the thermionic diode rectifier, for example, there is only one other electrode, the anode. Electrons can pass only from the heated cathode to the anode. Such valves ruled for about the first half of this century: the whole electronics industry, of which radio and television receivers are the most obvious signs, depended upon them. But since about 1950 semi-conductor devices have begun to replace them, and since about 1960 the number of thermionic valves manufactured annually has actually begun to fall, in spite of the continuing growth of the electronics industry. For many applications, as we shall see later in this chapter, only semi-conductor devices can be used, because they can be made small and packed compactly; thermionic valves are too hot, large and clumsy.

We have so far mentioned only the *diode*, and should consider also the *triode*, whose most characteristic use is in amplification. In the thermionic valve triode a third electrode in the form of a mesh, the control grid, is inserted between the cathode and the anode. This does not block the flow of electrons but effectively controls it. The total electron current collected by the anode is very sensitive to changes in the voltage applied to the grid – far more so than to changes in the voltage applied to the anode itself. This is the basic condition for amplification, where small changes in voltage at one point in a circuit can be made to cause large changes in voltage elsewhere.

The form of the triode in solid-state electronics will cause no surprise. In Fig. 28 (b) we illustrate the *junction transistor*. Here again a third connection is introduced, this time to the middle element of a 'sandwich'. A piece of n-type is sandwiched between two pieces of p-type. Again consider a positive charge passing from left to right. It will pass across boundaries A and B without difficulty, but will fail to cross boundary D because of the depletion of positive holes at the left of boundary D. But we can introduce holes at C by causing a positive charge to flow inwards at this boundary. The total current flowing from left to right is naturally highly sensitive to the size of this input current – again, the necessary condition for amplification.

The semi-conductor revolution faced many difficulties at first. Before impurities can be introduced under controlled conditions in minute proportions (say one in a million), the host substance has to be pure to an exceptional degree. Less than one impurity atom per 10^{10} host atoms can be tolerated. But also, in junction devices of the types described above, the two or three parts have to be made integrally in a single crystal; to prepare the sections separately and attempt to stick them together would not give sufficiently intimate contact. We shall not discuss the fascinating methods used to meet these formidable requirements, nor have we the space to describe the many other devices which depend upon the properties of semi-conductors, such as sensitive light detectors which make use of the property of *photoconductivity* – increased electrical conductivity under the action of light. We must leave the reader with this brief guide to the understanding of the new technology, and with a number of references for further reading in the Bibliography.

Superconductivity

The phenomenon of superconductivity occurs only at very low temperatures, each metal or alloy which can exhibit the property having its own characteristic transition temperature below which the electrical resistance suddenly disappears. Most of these transition temperatures lie between 1 and 10 K; the highest known is somewhat below 20 K. So for any practical use to be made of the property the temperature of the working material has to be lowered by the use of liquid helium. Until about twenty years ago this would have been possible in only a few laboratories in each of the scientifically developed countries. Now, the technology of *cryogenics* has advanced to the point at which anyone who wishes to study or use superconductivity can do so with a modest expenditure of money or effort. He can buy liquid helium in small quantities and have it delivered by rail in special containers, if he has no helium liquefier of his own.

Even so, the effort involved in working at liquid helium temperatures is not negligible, and makes some of the more obvious applications of superconductivity at least problematical, if not ultimately impossible. As we have already mentioned, the advantages of being able to carry electrical power without any loss in the transmission cables would be enormous. But the whole of the cable, made of one of the superconducting metals, would have to be maintained at temperatures in the range attainable with liquid helium – a somewhat terrifying undertaking.

For about fifty years after the discovery of superconductivity its use in electrical power transmission would certainly have been thought impossible, for a curious reason. A magnetic field destroys superconductivity, and this is true even of the magnetic field generated by the superconducting current itself. So only a certain limited electric current can be carried in any piece of superconducting material. Since about 1960, however, the situation has been changed by the discovery that certain substances, mainly alloys, can remain superconducting under the action of very high magnetic fields, and can therefore carry very high electric currents. It appears that in these materials a state is spontaneously established in which a very large number of fine filaments in the superconducting state alternate with filaments in the non-superconducting, or normal state. The current is of course carried by the thin superconducting filaments: it had been known for a long time that the *surface* layers of a piece of superconducting material could carry quite large currents even in the presence of a magnetic field. In the new alloys, Type II superconductors, as they are called, the whole of the superconducting part of the material seems to be able to take advantage of this because it is all near a surface, even though the surface is an 'internal' one separating the superconducting and normal filaments.

An application of outstanding importance made possible by the discovery of Type II superconductors is the *superconducting magnet*. This consists of a coil (or 'solenoid') of superconducting wire in which a current has been initiated, and once established, left as a 'persistent current'. Because there is no dissipation of heat by the current, a closely wound coil of very fine wires may be used, carrying a high current, and very high magnetic fields may be generated. The advantages are spectacular. If one requires the magnetic fields to be generated over a fairly small volume, say a few tens of cubic centimetres, the whole magnet can be contained in a Dewar vessel containing only a few litres of liquid helium. In such a piece of apparatus the same magnetic field could be generated as would require a large installation of the conventional type, comparable in scale with those used in the experiments of nuclear physics.

'Superconductivity' is not a household word, as is the word 'transistor', for example. High magnetic fields are not everyday requirements, and are more appreciated in the research laboratory than in the factory, or the home. But many other subtle applications of superconductivity have been exploited in the laboratory, and no doubt many are used in secret devices which have not been publicly described. Tiny devices using superconductors can replace semi-conductor devices for many applications, and have the advantage, of course, of even lower heat dissipation, so that they can be packed even more tightly. Experiments on the containment of hot plasma for the generation of controlled thermonuclear power, mentioned in Chapter 4, have used superconducting coils for the production of the magnetic field. But the real breakthrough into a major new technology has hardly occurred. The real prize would be won if a material were discovered which was superconducting at ordinary temperatures and did not require cooling. Even this may be possible! At least it is seriously considered by hard-headed scientists in certain industrial laboratories.

Computers

As one might suppose, the computer was invented to compute; in other words it is, essentially, a calculating machine. Its more dramatic characteristics are, in a sense, by-products of the techniques which are necessary for the performance of complex calculations. So we must begin by discussing the way in which calculations are made, taking as example the basic process of all calculation, that of addition.

Anyone who has used, or seen in use, a calculating machine such as a cash-adding machine in a supermarket, or even an *abacus*, will be able to understand the elementary process of the computer. The adding machine or the abacus simply performs rapidly, by the mechanical action of the cog-wheels or by the direct action of the finger, the same operations as are performed when one adds columns of figures on paper. It is worth being reminded what some of these operations are.

If we wish to add the following figures: 80, the first operation
                                    22
                                    48

is to *add* the figures in the units column, making 10. We find this
easy because we are so used to adding small numbers; if in doubt,
however, we could find some simple way of proving it – perhaps
using an abacus! We now *write* 0 in the units column, and *carry*
the 1 (which of course now means ten) to the tens column, thus:

$$
\begin{array}{r}
1 \\
22 \\
80 \\
48 \\
\hline
0
\end{array}
$$

Performing the same operation in the tens column, we find that
the figures now in the column, added, make 15, and the addition
is complete:

$$
\begin{array}{r}
22 \\
80 \\
48 \\
\hline
150
\end{array}
$$

So we see the necessity to be able to add small numbers, and
to be able to write down figures in particular positions, or to
carry them to other positions. In the abacus the carrying is
done by the finger, in the calculating machine by the automatic
action of cog-wheels.

But what is it in the computer which corresponds to the
written figure? The answer is perhaps unexpected, and depends
upon some of the developments of solid-state physics discussed
in the previous chapter. The most common is the direction of
magnetization of a tiny ring of magnetic material. Just as the
common bar magnet may be magnetized either in one direction
or another, so may a ring be magnetized either in a clockwise or
an anti-clockwise direction. A current in a wire passing through
the centre of such a ring will magnetize the ring in one of these
directions; passing in the opposite direction it will magnetize the
ring in the opposite direction. So we see the possibility of 'writing

down' figures. Then if the wires are suitably connected to others with other components such as transistors (acting essentially as switches), the information stored in one ring can be 'carried' to another. In modern computers there may be *stores* consisting of thousands, or even millions of such rings, each one carrying its bit of information. We have mentioned in the previous chapter some of the solid-state devices of modern electronics, such as the transistor, which allow smaller and smaller circuits to perform more and more complex operations. To meet the needs of computer technology the art of 'miniaturization' has been pressed to the point at which a complete circuit, including perhaps transistors, connecting leads, resistors and capacitors, can be built into a single piece of semi-conductor. This is the world of *integrated circuits*, or *microelectronics* (see Plate XII).

But the single ring can exist in only two magnetic states, clockwise or anti-clockwise, and this surely cannot be made to represent more than two basic figures, whereas in the calculating machine there are ten (0, 1, 2, 3, 4, 5, 6, 7, 8, 9). The fact is, that in spite of this limitation, the enormous speeds of the operations which depend upon electrons rather than upon cog-wheels (or beads), make it worth while to operate in a system in which there *are* only two basic figures, say 0 and 1. This is the *binary* system.

Turning back to the decimal system for a moment, we know that 150, for instance, means: $(1 \times 100) + (5 \times 10) + (0 \times 1)$. In the decimal system, therefore, we deal in powers of ten (1, 10, 100, 1000 etc.). Now in the binary system we have to deal in powers of two (1, 2, 4, 8 etc.). Instead of having columns for units, tens, hundreds, etc., we have columns for units, twos, fours, etc.

Then

0 in the binary system is the same as 0 in the decimal system

1	,,	,,	,,	1	,,	,,
10	,,	,,	,,	2	,,	,,
11	,,	,,	,,	3	,,	,,
100	,,	,,	,,	4	,,	,,
101	,,	,,	,,	5	,,	,,
110	,,	,,	,,	6	,,	,,
111	,,	,,	,,	7	,,	,,

and so on.

With a little practice the reader will be able to develop this further, and also the rules for addition, that

$$0 + 0 = 0, \qquad 0 + 1 = 1, \qquad 1 + 1 = 10$$
and so on.

Then the addition performed above in the decimal system would look like this in the binary system:

$$
\begin{array}{ll}
10110 & (22) \\
1010000 & (80) \\
110000 & (48) \\
\hline
10010110 & (150)
\end{array}
$$

Obviously we cannot develop the whole science of computing in these pages. We can, however, pick out the basic principles and sketch in enough of the science to convince the reader of its plausibility. He will understand that a pulse of electricity, which of course could magnetize (or demagnetize) a ring, could represent the digit 1, while the absence of a pulse could represent the digit 0. Graphically, the pulse could be illustrated thus: $\sqcap$, while the absence of a pulse could be illustrated thus: ＿. The number 22 in the decimal system, or 10110 in the binary system, would be: $\sqcap\sqcap\sqcap$. A series of pulses flowing in suitably designed circuits could be made to 'add', 'write', or 'carry' figures.

Once the operation of addition can be carried out, the whole of mathematics can be developed. Multiplication, for example, is only multiple addition ($5 \times 22 = 22 + 22 + 22 + 22 + 22$).

So fast is the passage of electricity that the elementary arithmetical operations can be performed in times of the order of millionths of a second. In this fact lies the great power of the electronic computer. Indeed, the time taken by the calculation itself is usually negligible in comparison with the time taken to feed in the instruction – or 'programme' – and to extract the answer. The science of 'programming' a computer is a relatively new one, but already consumes a substantial proportion of our mathematically trained graduates and technicians. Of course, every user of a computer does not have to be so highly trained. For a given purpose he can use a programme, prepared by an expert, which converts his own instructions – which may literally be typed instructions – into a form which can be accepted by the computer. The computer can then operate upon data fed into it or upon data already stored in it, according to these instructions. One of the first things it must do will be to convert instructions or data as necessary into the binary system; one of the last things will be to convert the answer from the binary system into an acceptable form. We have, of course, been describing the electronic *digital* computer: it is not the only type of computer, but by far the most important.

We have not yet explained why computers can be so important in industry and commerce, or even in ordinary office routine. But in fact much of the routine work in commerce is mathematical, though of a somewhat low order. A computer could, for example, be instructed to multiply the hours worked by all employees by the basic rate and then to deduct the appropriate tax. Another great consumer of office time is the search for records. But the computer can be made to do this, and indeed, to hold the records in its store. In banking, the numbers printed on cheques with magnetic ink can be 'read' by the computer, which will then 'find' in its own store the customer's account and carry out the necessary operations upon it. In the office of an airline a customer's enquiry about a possible reservation will be typed in his presence, transmitting a message to a central computer which has in its store a record of the state of bookings for all flights. The answer comes back immediately, and is typed automatically.

Of course, it is not quite so magic as it looks. The booking clerk has typed the message in an acceptable form, that is, in a conventional 'language', using only words which the computer recognizes. The computer replies in the same language. One can only instruct a computer to do what it knows how to do; that is, what it has been programmed to do. But the possibilities are almost limitless, provided one is able to develop a suitably complex programme, and provided the computer is large enough and fast enough. They certainly include, in principle, the possibility of translating by computer from one (human) language into another! A small computer for simple applications may cost a few thousand pounds; the largest computers cost millions of pounds.

We have mentioned the instructions given by typewriter or by magnetic ink. But there are many other media through which data or instructions may be fed to the computer. It may 'read' slots punched in card, or holes in paper tape, or records stored on ordinary magnetic tape. Again, the 'answer' may be presented by typewriter or other printing device. But the answer may be translated into immediate action through the operation of electromagnetic relays or other control devices. A possible instruction to an office computer might be to pick out all the cards in a stack which bear a certain combination of data (perhaps all the physicists with a Ph.D. degree above a certain height who can speak Russian). We have come a long way from the abacus.

Lasers

There has not so far been much reference to light waves in this book, except for their somewhat academic significance in the development of the fundamental theories of modern physics. We also mentioned briefly the earlier synthesis in physics which had showed, by the end of the last century, that light waves were, so to speak, a consequence of the laws of electromagnetism. For many centuries the laws of optics, summarizing the behaviour of light in vacuum, air and other transparent media, have been at the centre of classical physics. All students of physics have learnt and applied these laws; devices from the simple spectacle lens to the modern 'zoom' lens of film and television have depended upon them.

In the preceding chapter the similarity in nature between ordinary visible light and other forms of electromagnetic radiation – such as radio waves – has been described. But there is an important fundamental difference between them. In the radio waves sent out by a transmitting aerial, say, the whole of the energy is 'in step'. The radio wave is generated by a single alternating current flowing in the aerial. On arrival at the aerial of the receiver it generates another alternating current, which is then amplified enormously through many stages to produce, ultimately, the sound or television signal. But ordinary light is generated, as we have seen, by atomic 'oscillators', or, more

161

precisely, by the transitions of electrons between different atomic quantum levels. And the atoms are not in step! In a sense we see the light emitted by millions of tiny aerials, all oscillating at random. Even if, like the radio wave, the light is *monochromatic*, that is, it has only one colour, or wavelength, the waves are still out of step with each other. In the accepted terminology, the radio wave is *coherent*; the light wave is *incoherent*.

But there can obviously be no reason, in principle, why waves having the wavelengths of visible light should not also be coherent, if the atomic 'oscillators' could be made to 'oscillate' in step with each other. This is what happens in the laser.

In the preceding chapter we referred to the fact that atoms, or molecules, or indeed, the whole of a lump of solid, have series of quantum levels which the electrons must occupy. Incident light (or photons) can cause electrons to jump to higher quantum levels, when the photon is *absorbed*. Conversely, an electron falling to a lower level causes the *emission* of a photon. The emission of light by matter is thus due to 'downward' transitions from higher to lower quantum states, following excitation to the higher state by some means – either by collisions between atoms, as in a hot body, or by the previous absorption of light.

Now only a small number of the atoms in matter are normally excited, that is, have any of their electrons in excited states. But under certain special circumstances a situation known as 'population inversion' can be created, in which there are more atoms in high excited states than in normal states. In these circumstances a photon passing through the matter – either coming in from outside or originating as a result of a downward electronic transition inside the matter – can cause an excited atom to emit a photon. The two atoms can then cause other atoms to emit photons and eventually an 'avalanche' of photons is emitted. Of course, because light travels at an enormous velocity this takes only a very short time. The important points for us are these: all the photons emitted correspond to the same wavelength of light, and the light is both closely *parallel* and closely *coherent*.

'Laser' stands for 'Light Amplification by Stimulated Emission of Radiation' – a self-explanatory title in the terms of what has been said.

Of course, the state of population inversion is not a natural one. But immediately after the absorption by matter of a bright flash of light of suitable wavelengths it can exist for a very short time – until the atoms have returned to their normal dispositions. A typical – though not the only – way of inducing laser action makes use of such bright pulses of incident light. Under correct conditions another, intense pulse of 'laser light' is then emitted, having the special properties of being parallel and coherent. By the use of clever devices, which we shall not describe here, the light emitted can be made to 'bunch' into an even more intense and shorter pulse.

It will be hard for the non-scientist to appreciate why the possibility of obtaining and using beams of monochromatic, closely parallel and coherent light excites the scientist. Since 1960, when the first laser was operated successfully, millions of pounds have been spent annually on research into lasers, and thousands of scientists have been involved. And yet the uses of lasers so far have seemed to be rather trivial. Most of them make use of the extreme parallelism of the light, which can be concentrated by lenses on to very small areas. An ordinary 'ruby' laser (one using ruby as the working material) can concentrate light energy at the rate of as much as 100 kilowatts upon an area of the size of a pin-head – many thousands of times brighter than the brightest image of the sun which could be focused upon the same area!

We have to qualify the above statement in two respects – both favourable to the possibility of developing uses for the laser. First, in lasers of the type described above the energy of the beam comes out in pulses. It was correct to use the unit kilowatt above, because this measures the *rate* of production or consumption of energy – as every householder has to understand. But the figure given was the *average* rate, and because the light is concentrated into short pulses, each lasting perhaps only a thousandth of a second, the rate of production of light energy *during* the pulse is higher still. Secondly, because the light is monochromatic, it is still brighter, as compared, say, with the image of the sun, at its own sharply defined wavelength.

Is the laser beam, because of its intensity, and its extreme parallelism, the 'death ray' of science fiction? Does this account for the scale of the effort which has been put into the study of lasers by military research establishments? It seems to be unlikely. The laser is an inefficient device in that only a small fraction of the energy fed into it comes out in the beam; moreover, laser light, like ordinary light, is much affected by atmospheric conditions such as cloud, or fog. It has already been possible – though only just possible – to detect laser light reflected from the surface of the moon; because of the parallelism of the original beam sent out from the Earth, the area of the moon illuminated would be quite small. In principle, very fine detail on the moon's surface could be 'seen' by this method. However, the transmission of laser light *outside* the atmosphere might have dramatic applications. A laser beam is so strictly parallel that a beam sent out from a satellite in orbit around the Earth could be directed anywhere else in the solar system, and detected there. One possible application of the laser might be to initiate thermonuclear reactions, as mentioned in Chapter 4.

On a much smaller scale the laser has already had uses which depend upon the intensity or parallelism of the beam. It can be used for 'punching' tiny holes in diamonds. Naturally, a laser beam can be very dangerous, especially to the eye. But its action upon tissue has been put to use. It has been used successfully by eye surgeons – though of course at very much lower intensities – for 'microwelding' detached retinas.

The other class of applications of lasers depend upon the coherence and monochromatic nature of the beam. Potentially, there would be enormous scope in the field of communications if laser light waves could be used and handled in the same way as their coherent and monochromatic brothers of much longer wavelength, the radio waves. In principle this can be done; the difficulties – formidable ones – lie in developing practical techniques and, again, in the limited atmospheric range of visible light. But because of the much higher frequency of light waves than radio waves, much more information could be carried. A single laser beam could carry millions of telephone conversations, or radio or television programmes; the ordinary radio beam normally carries one or two.

A curious and interesting application in the same class is the *hologram*, a device with which all the information contained in a three-dimensional 'photograph' may be put on a two-dimensional photographic plate. When the plate is suitably illuminated by laser light of the same wavelength as that of the laser originally used in taking the photograph, the viewer sees what appears to be a three-dimensional image behind the plate. Here the light must not be pulsed, and a 'gas laser', which gives a continuous output, must be used. We shall not explain this device further, but remind the reader that a somewhat – though not quite – analogous situation has been mentioned earlier. The X-ray structure analyst investigates a three-dimensional solid, and secures on a two-dimensional plate a great deal of information about the kind of lattice arrangement upon which the atoms are arranged. X-rays are used because their wavelengths are about the same as the distances between atomic centres. In the hologram, on the other hand, laser light gives the ordinary 'visible' likeness of the object.

The development of the laser is a good illustration of the way in which techniques of the utmost sophistication, based on scientific discoveries of great subtlety, have burgeoned in the new scientific revolution. Millions have been spent; thousands have worked on it. We have not set out the details in this chapter (why, for example, is 'ruby' used?) precisely because they are so sophisticated and subtle. It is, however, a particularly interesting slice of post-war science. For certain highly specialized purposes, particularly in research, the laser is unique. For example, a modern repetition of the Michelson–Morley experiment, using laser light, has confirmed the null result to about a thousand times greater accuracy. But for many of the much publicized applications of lasers there are cheaper – and easier – ways of doing the same thing.

CHAPTER 8

The New Biology

In the Introduction to this book we mentioned briefly the unity of biology with physics and chemistry. Recent discoveries have welded them still closer together. The reader has not been given here the background which would allow him to appreciate fully the place of biology in the new scientific revolution. Nor, indeed, are the authors competent to set it out. It is still true, however, that some of the discoveries in biology made since the Second World War have been among the most striking and exciting in the whole history of science, and ought not to be ignored completely, even if we were writing a book only about *matter*. In this short chapter we illustrate the point by mentioning some of the most important. In nearly every one the link with physics and chemistry has been fundamental in two ways: first, the discoveries have required the use of the most advanced experimental techniques of modern physics and chemistry. Secondly, the processes which have been studied have been shown to *be* physical and chemical processes. They have of course been concerned with highly specialized areas of physics and chemistry, particularly the properties of certain giant molecules, which would not have been so thoroughly cultivated if they had not held the secrets of biological action.

One link between biology and physics is known to every schoolboy: the fact that the muscles of a frog can be made to twitch when an electromotive force (or electrical 'voltage') is applied to it. A fact which is, in a sense, the converse of this, has been known for over a hundred years: that nerves and muscles themselves generate electrical pulses. But now, as a result of work carried out mainly during the last twenty years, it is known that the pulse *is* the signal which is propagated – the nerve impulse itself. And it is propagated along parts of the nerve fibre which are remarkably reminiscent of insulated submarine cables. Other parts of the fibre take part in re-inforcing the nerve signal – perhaps reminiscent of the amplifiers which are introduced at intervals in such cables. The study of the nerve cell itself has reached the point where it is known at what part of the cell signals are received, where they are 'sifted', and where they are passed on to the 'submarine cable'. To be able thus to glimpse simplicity while surrounded by complexity is the first aim of the scientist. But as in all the fundamental problems of biology, the complexity – and the problems to come – are formidable.

Nearly all the chemical substances which are important in biology are compounds of carbon, and very many include carbon chains. It used to be thought that the chemical compounds derived from plants and animals – *organic* compounds, as they were called – could not be created in the laboratory. It was believed that a mysterious barrier existed between these substances and the *inorganic* compounds derived mainly from mineral sources. But about 150 years ago it was discovered that the organic compound *urea* (found in animal excretions) could be formed by heating ammonium cyanate. Although of course this substance also contains carbon, it would have been regarded as an inorganic substance. The lesson has been thoroughly learnt. No scientist now believes that any mysterious intervention is necessary before an organic compound can be prepared in the laboratory. All that is required is back-breaking work! Many of the most remarkable events in biological science have been the syntheses of organic substances of medical importance – of which *insulin* is perhaps the prime example.

But more recently, determinations of the *structures* of even more complex substances have been at the centre of biology. The techniques used have included some which we have already discussed – particularly those of X-ray analysis – carried to the ultimate level of refinement. We shall mention one: Perutz's extraordinary determination of the structure of *haemoglobin* – an essential ingredient of red blood cells. What we mean by the

167

determination of a structure is, of course, that the atoms of the molecule have to be identified and their positions determined in relation to each other. For example, the molecule of the very simple substance ethane has the structure

$$
\begin{array}{ccc}
\text{H} & \text{H} \\
| & | \\
\text{H}-\text{C}-\text{C}-\text{H} \\
| & | \\
\text{H} & \text{H}
\end{array}
$$

This has 8 atoms in the molecule, is made up of 2 elements, carbon and hydrogen, and its 'molecular weight' is 30. It can be regarded as having the form of a very short chain, two atoms long, with simple attachments.

Now haemoglobin has about 10 000 atoms in the molecule, its molecular weight is 64 500, and six elements are involved. (It is remarkable how few of the 92 naturally occurring elements enter into organic chemistry.) The X-ray diffraction patterns contained tens of thousands of spots, their analysis required months of computer 'time', and the whole study – which, in spite of the above remarks, depended upon insights rather than upon techniques – took about 16 years. Even now it cannot be claimed that the position of every atom in the molecule is known. But the general *shape* of the molecule is known. It is a very long, complex interwoven chain in the form of a helix, with short attached 'arms', made up of hundreds of identifiable 'sub-molecules' or *radicals*, and more than half of these arms have been clearly identified.

In Perutz's research unit at Cambridge was also made perhaps the most exciting discovery of the period – that known popularly as the breaking of the 'genetic code' – the explanation of heredity in physico-chemical terms. Although hundreds of scientists had been involved in the slow approach to the discovery, the final and decisive insight was due to Watson and Crick.

It has been known for a long time that heredity is controlled in the *chromosomes* of biological cells. Before a cell divides into two (the process of *mitosis*), chromosomes appear in the cell nucleus, duplicate themselves, and move to opposite ends of the nucleus. The two new cells are then copies of the original single cell. Chemically, the main substances in the nucleus are *nucleic acids*, which have very large, complicated molecules. One of these, *deoxyribonucleic acid* (known as DNA), carries the heredity of the cell. The problem was to explain how molecules of this substance could 'carry' the genetic information, and be self-replicating.

In 1953 Watson and Crick announced their 'model' of the process. Now DNA is not, like simple substances, made up of a single type of molecule, but may vary in structure within certain limits. Again, the molecule is a very long chain (thousands of atoms long), again with short 'arms' along its length, and there are four kinds of arm, known by the letters A, G, C, and T (the first letters of their chemical names). Their answer to the above question was as follows. The molecule is a *double* helix in which the two strands are 'stuck' together through these arms. But G will only stick to C, and vice versa, and A will only stick to T, and vice versa. The 'coded' information is determined by the particular sequence of the arms A, G, C, T along the length of the helix. The self-replication is imagined to begin by a separation of the two helices. At once they begin to build up their complements from 'spare' matter in the cell. Every A finds a T, every G finds a C, and ultimately there are *two* double helices, each a copy of the original. Cell replication, and therefore heredity, are explained.

The general correctness of this explanation is accepted. Again, as biology has become conceptually simpler, the complexity and formidable nature of the problems to come are more clearly recognized. But a most remarkable feature has now entered our description, in that we have attributed to the location of certain chemical combinations of atoms what we identify as 'information'. This is surely reminiscent of our discussion of the digital computer, where the magnetic dispositions of numbers of tiny rings also carried information.

This kind of thinking begins to pervade biology. In particular it is recognized that the most complex of biological systems, the *brain*, is a kind of computer. Like the electronic computer, it is a network of 'switches', although in the brain these are of course nerve cells, not transistors. But it is an enormous computer,

169

having perhaps ten thousand times as many 'switches' as the largest electronic computer. The inter-connections, too, are far more complex. While each switch in a computer is connected to only two or three others, each one in the brain is probably connected to hundreds.

A large computer can seem to reason, and to make decisions, and is of course incomparably more powerful than the brain in some situations. However, the much 'larger' brain is incomparably better in others. Its 'store', or memory, as we are used to calling it, seems to be more efficiently laid out. It can choose what to store more sensibly, and is better at finding what it wants in the store. It is better at 'learning'; it also seems to be able to adapt, or 're-wire' itself.

But in a sense, the brain is a computer; memory, subconscious memory, thinking, even emotion, all have a physico-chemical basis. Should this worry us, each of us with his feeling of uniqueness and self-awareness, of being in control of himself and not 'programmed'? It is surely an unreal question. If we feel we have free-will, we have it; if we feel we have an emotion, we have it. The fact that even highly personal and private thoughts and feelings are physico-chemically based makes no difference to one who does not know it; how can it make any difference to one who does?

CHAPTER 9

The Solar System

The Solar Constant

For us living on the earth the most important other single large body in the universe is the sun. The sun is our main source of light, heat and energy. Of course, there are some energy sources on the earth, such as earthquakes, volcanoes, hot springs and lunar tides, that do not come from the sun. Nor is nuclear energy due to solar action. But the main energy sources which we use at present in everyday life, like gas and electricity, are obtained from coal, oil and water-power; the energy in coal and oil is ultimately derived from solar energy trapped by plants in the far distant past, and that in water-power is due to solar energy which raises vapour from the seas and oceans. These sources, as well as winds and solar tides, all bring energy to us which can ultimately be traced to the action of the sun on terrestrial matter.

What then is the sun? It can be regarded as a machine which generates its own energy as it radiates. As far as we on earth are concerned, its most important feature is its regularity of behaviour. Not only does the sun radiate energy at a steady rate, but it also holds the earth at a more or less constant distance. We tend to take all this for granted, for, as an American astronomer has remarked, the sun is like a dependable and steady husband who is rarely fully appreciated! If the sun's action on the earth were not so remarkably steady, there is no doubt that life would soon perish from the earth.

The steadiness of the sun's supply of energy to the earth is expressed in terms of what is known as the 'solar constant'. This is defined as the amount of heat in calories which would fall in one minute on an area one square centimetre placed perpendicular to the radiation as it falls on the earth's surface, if the earth had no atmosphere and were at its mean distance from the sun. Consequently, to determine the solar constant, the actual amount of energy which falls per square centimetre per minute must be measured and then the result must be corrected for the absorption by the earth's atmosphere. The final figure is nearly two calorics per square centimetre per minute.

From the ages of fossil-bearing rocks which have been determined by means of radioactive measurements, we deduce that life has existed on the earth for upwards of 600 million years. During all this time the solar constant cannot have changed greatly, because if it had become either twice as large or half as large as it is now, then, in either case, all life would certainly have perished. Consequently, both the solar distance and the sun's rate of generation of radiation must have varied comparatively little over this vast span of time, which, on some theories at least, may be between 5 and 10 per cent of the age of the whole universe.

The Sun's Distance

The sun's distance from us, although on the whole very nearly constant, oscillates a little. Its average value is taken as the fundamental distance in astronomy and is known as the astronomical unit. There are several methods by which it may be estimated. For example, it may be determined by means of the solar parallax, which is the angle subtended by the radius of the earth at the centre of the sun; knowing the radius of the earth, we can then calculate the distance. In practice, the solar parallax is obtained by considering the position of the sun relative to the background of the stars from two widely separated observatories on the earth's surface. Knowing the base-line between these observatories and measuring the relevant angles, the solar parallax, and hence the sun's distance, can be determined geometrically. The result thus obtained can be checked as there are other methods by which this fundamental unit may be estimated.

Indeed, the sun's distance from the earth can be calculated indirectly by determining the distance of any one planet. This is a consequence of the remarkable fact that, without measuring any celestial distance whatsoever, we can draw a scale-map of the solar system, although of course we do not know, without measuring some distance, what the scale is. This can be done in more than one way. The most important method depends upon Kepler's law of planetary motion, that, to a first approximation, the squares of the periods of revolution of the planets around the sun are proportional to the cubes of their mean distances from the sun. From the angular motions of the planets their periods can be obtained and the ratios of their mean distances determined. But one distance, say the distance of a particular planet from the earth or from the sun, must be measured in order to determine the scale and hence the actual distances.

The most recent determinations of this kind were based on observations of one of the minor planets, Eros, which, in 1931, came within 16 000 000 miles of the earth, which is a small distance for celestial objects. As a result of many years work on these observations, in which the leading part was taken by Sir Harold Spencer Jones, the mean distance of the sun was ultimately determined to be 93 009 000 miles.

A new method of measuring the solar parallax has been developed with the aid of radar. Echoes have been detected of radio pulses transmitted to Venus from Jodrell Bank and from a station in the United States, and the time of travel to and from the planet (about five minutes) has been determined. The experiment has already given a new measurement of the solar parallax which corresponds to a mean distance of the sun from the earth in good agreement with the value obtained optically.

The sun's distance, however, is not absolutely constant. This can immediately be deduced from the fact that the sun's diameter does not always subtend the same angle at the earth. During the course of the year, this angle varies by more than one minute of arc. Although the seasons are not primarily determined by this variation in distance, but by the fact that the earth's axis of daily rotation is not exactly perpendicular to the earth's orbit, it has an important secondary effect. In December the sun is only about 91 000 000 miles away and in June about 94 000 000 miles; consequently there are greater extremes of climate in the southern hemisphere.

The Rate of Energy Generation by the Sun

Knowing the sun's distance, it is possible to calculate the size of the sun and hence, given the solar constant, the rate of emission of energy per unit of surface area. Its diameter is found to be about 864 000 miles, or about 109 times the diameter of the earth. Each square centimetre of the solar surface radiates, on the average, about 90 000 calories a minute, i.e. at about the rate of a 7 kilowatt engine. The earth receives rather less than one part in 200 million of the total energy radiated by the sun, but even this small fraction is something like 4 million kilowatts per square mile. We have here an immense potential source of energy which has not yet been effectively utilized.

In the course of the year the total output of heat from the sun amounts to 3×10^{33} calories. This seems an enormous quantity, but to appreciate its significance we must relate it to the mass of, or amount of matter in, the sun. This can be determined by appeal to the Newtonian theory of gravitation. Knowing the radius of the earth, the gravitational acceleration on the earth's surface, the distance of the sun and the acceleration of the earth in its orbital motion, we can compare the mass of the sun with the mass of the earth: the sun is about 332 000 times as massive. To obtain the mass of the earth, the constant of gravitation which controls the strength of gravitational fields must be determined experimentally. An accurate value was obtained more than three quarters of a century ago – a fairly good estimate had already been obtained in the eighteenth century by Cavendish – from which it was deduced that the mass of the earth is about 6×10^{27} grammes. Consequently, the mass of the sun is about 2×10^{33} grammes.

Comparing this figure with the sun's energy output per year of about 3×10^{33} calories, we see that to each 2 grammes of the sun's mass there is a generation of energy of 3 calories per year. This does not seem very remarkable at first sight, but it becomes utterly fantastic when we take into account the enormous time for which it has continued. From the evidence of past life on the earth, it is clear that this steady generation of energy must have been going on for upwards of 600 million years and probably or considerably longer. Thus, during 600 million years or more, nearly 1 000 million calories have been generated *per gramme* of the sun's mass at a more or less steady rate. This is a truly stupendous figure. It has been calculated that 1 gramme of the ideal mixture of coal and oxygen for complete combustion could generate about 2 000 calories, so that, if the sun were composed of such an ideal mixture of coal and oxygen, the whole of the sun's mass would have been burned up in rather less than two thousand years. Thus, chemical combustion is obviously not the method by which the sun generates energy.

During the latter part of the nineteenth century a great deal of attention was paid to this problem. It was one of the great puzzles of physical science at that time. The German physicist Helmholtz looked for some other source of energy generation and came to the conclusion that the only source which could endure for such a long period was gravitational contraction. He pointed out that, if the sun slowly contracted under its own gravitation, there would be a loss of gravitational energy which might be a source of radiant energy. To generate sufficient energy the sun's radius would have to contract at the rate of 75 metres per year at the present time. If we calculate backwards to get a steady rate of energy generation all the time, we find that the rate of contraction would have been faster in the past. Extrapolating to a state when the sun was a very tenuous nebula, the longest period that can be obtained is approximately 20 million years. Helmholtz thought that this was roughly the age of the sun, but he and his followers were involved in a notorious controversy with geologists who rejected this period as utterly inadequate, although at that time they had no precise means of dating fossils. The controversy was only resolved this century with the aid of nuclear physics. Before going into this problem further, two other questions must be considered: first, the sun's temperature, and second, the sun's chemical composition.

The Temperature and Chemical Composition of the Sun

Knowing the solar constant, it is comparatively easy to estimate the surface temperature of the sun. We can appeal to different physical laws characterizing so-called black-body radiation.

175

(A black-body is a perfect radiator – see, for example, Fig. 27.) According to which law is invoked, temperatures of 5 200 and 5 600 K, respectively, are obtained. The fact that these two are not exactly the same shows that the sun is not quite a perfect radiator, but the agreement is sufficiently good to show that the sun's effective surface temperature must be in this region. These figures may be compared with the temperature of an ordinary electric arc which is about 4 000 K. They may sound large by our standards, but are very modest for celestial temperatures. Theories of stellar structure enable estimates to be made for the temperature at the centre of the sun. It is generally accepted today that this temperature must be of the order of 12 000 000°C.

The second question that must be considered is the composition of the sun. More than a century ago, the famous French positivist philosopher Auguste Comte stated dogmatically that we shall never know of what the stars are made. He was answered some years later by the rise of the science of astronomical spectroscopy, by means of which the chemical composition of any incandescent celestial body can be analysed provided a sufficiently detailed spectrum is obtained. The sun's spectrum is found to be enormously complex; it is a continuous spectrum crossed by a large number of dark lines called absorption lines. The continuous spectrum is produced by the photosphere, or surface of the main body of the sun. Above this is the lower part of the solar atmosphere, known as the reversing layer, which, being at a somewhat lower temperature, absorbs from the continuous spectrum below the particular colours, or spectral lines, which it would otherwise emit. Normally, the spectrum of an incandescent gas is an emission spectrum consisting of bright lines depending on the chemical composition. In the case of the sun roughly 20 000 absorption lines have been observed and more than half of them have been identified.

Broadly speaking, the same elements exist in the sun as here on earth. For various reasons some elements found on earth are not directly observed in the sun and in one remarkable instance an element was first detected in the sun by Lockyer in 1868 before it was discovered in the laboratory by Ramsay in 1895. This element was appropriately named helium.

To find the particular method by which the sun generates energy, it is essential to know not merely the actual elements that occur in the sun but their relative abundances. These can be deduced by studying the forms and strengths of lines in the solar spectrum. Although the proportions of metals, such as iron, nickel and calcium, are roughly the same on the earth and the sun, there is an important difference: the sun's atmosphere consists mainly of hydrogen, and to a much lesser extent of helium. This difference is not due to a peculiarity of the sun's composition, but of the earth's, for throughout the universe hydrogen is by far the most abundant element. It is thought to characterize the sun as a whole, not merely its atmosphere. This conclusion is the key to the solution of the problem of solar energy generation.

The Method of Energy Generation by the Sun

According to Einstein's mass-energy relation (see Chapter 6), there is bottled up in each gramme of matter about 20 million million calories of rest-mass energy. The sun's radiation, which has been going on for upward of 600 million years, has required the generation of nearly 1 000 million calories per gramme. If this figure is compared with the 20 million million calories locked up in each gramme of matter, it is clear that only one part in 20 000 of the total inertial energy of the sun need be drawn upon. Even if the sun has gone on radiating much longer than 600 million years this source would be more than ample.

How can this source be tapped? As already explained in Chapter 4, two possible processes are now envisaged, both of which require very high temperatures for their operation. They both involve the transmutation of hydrogen into helium. It will be remembered that the release of energy in this transmutation depends on the fact that, although the nucleus of the helium atom can be formed from the nuclei of four hydrogen atoms, the mass of the helium nucleus is actually slightly less than that of the four original hydrogen nuclei, the loss being about 0·8 per cent. This means that for every gramme of hydrogen the resulting loss of mass is equivalent to the release of roughly 160 thousand million calories.

177

Of the two processes, the one known as the hydrogen chain is believed to be the predominant reaction in the sun, while in larger stars the carbon cycle is thought to be the chief source of energy. We have seen that, in order to maintain a steady supply of radiation for about 600 million years, the sun need have drawn on only about one part in 20 000 of its available inertial energy, and in 1 000 million years it need have consumed the equivalent of only one-part in 10 000 of its mass. Since the loss of mass when hydrogen is converted into helium is about 1 per cent, only about 1 or 2 per cent of the sun's hydrogen need have been converted into helium in the past 1 000 million years. Consequently, over the vast range of time the net change in the sun's composition and constitution has been sufficiently small to account for the presumed regularity in its rate of radiation, thus showing that the theory is self-consistent.

The tremendous scale on which the solar furnace operates is evident from the fact that some 800 million tons of hydrogen are converted into helium every second. This results in a loss of mass exceeding 6 million tons a second which is converted into energy. It has been estimated that the sun can go on steadily burning up its hydrogen at this rate for another 50 000 million years, but in all probability the sun cannot consume in this way more than a fraction of its supply of hydrogen. In this case, the process may continue for only about 5 000 million years.

How does the heat produced inside the sun escape to the surface and so ultimately pass across the intervening space to us? It might be conveyed from the interior to the surface by conduction, convection or radiation. Despite the enormous pressures which must prevail inside the sun, the sun is believed to be gaseous throughout. Consequently, heat must be conveyed mainly by convection or radiation. Near the surface the solar prominences provide dramatic evidence of vertical mixing. Nevertheless, despite this evidence of convection, it is believed that radiation is responsible for most of the energy-transport from the interior of the sun outwards.

Solar Neutrinos

The hydrogen chain reaction involves not only the production of positrons but also of neutrinos (see Chapter 1). Since the neutrino does not take part in 'strong' reactions, nearly all those generated near the centre of a star such as the sun escape into space, and it is thought that between 2 and 6 per cent of the sun's energy is lost in this way. Recently, it has been realized that neutrinos emitted by the sun should, in principle, be detectable on earth, and an experiment has been performed for this purpose in a large tank, containing some four hundred thousand litres of cleaning fluid, placed a mile underground in a gold mine. (The reason why it has been performed at this depth is to minimize the production of neutrinos by cosmic rays.) The solar neutrinos are absorbed by the heavy istope of chlorine through the reaction

$$^{37}Cl + \nu \rightarrow {}^{37}A + e^-.$$

The radioactive argon produced in this way is then separated chemically from the fluid and the number of radioactive atoms produced is counted by observing the reverse reaction

$$^{37}A \rightarrow {}^{37}Cl + e^+ + \nu.$$

The great significance of this experiment is that it provides us with direct information about conditions near the centre of the sun, whereas the photons we receive from the sun come only from the surface layers. The higher the energy of a neutrino the greater the probability of its detection. The energy depends very strongly on the temperature, and we therefore now have the possibility of a direct check on the temperature at the centre of the sun which we can compare with the value that has been obtained theoretically.

The surprising result of the underground solar neutrino experiment is that no neutrinos have yet been detected, although it should have been possible to observe a neutrino flux of only about one seventh of that predicted theoretically. The reason for this discrepancy is not yet clear, but so far astronomers have tended to think that it is more likely to be due to uncertainties in our knowledge of conditions in the interior of the sun, such as its chemical composition, rather than to some fundamental defect in the basic theory of stellar structure.

Solar Activity

Detailed study of the surface of the sun is facilitated by an instrument known as the spectroheliograph. The solar image is permitted to drift across the slit of a spectrograph, so that with the aid of a second slit at the focus the light passes to a photographic plate. In this way a monochromatic picture of the whole disc of the sun is obtained, that is a picture of the whole disc taken in one wavelength of light. This looks quite different from the sun as normally seen in light of all wavelengths simultaneously. Remarkable details are revealed, particularly if these monochromatic pictures are taken in the wavelengths of the most prominent lines on the sun's spectrum, pointing to the great turbulence of the solar atmosphere (see Plate XIII).

The best-known markings on the sun's disc are the great sunspots which were first realized to be solar by Galileo when, in 1610, he applied the telescope to astronomical observation. As a result of this discovery it was found that the sun rotates, for although the spots are not absolutely fixed on the sun's surface they can be used statistically to estimate the sun's rotation. This rotation is very different from that of the earth, which is the same everywhere. In the case of the sun, rotation varies according to latitude: near the solar equator the period is about $24\frac{1}{2}$ days, whereas near the poles it is about 34 days. The cause of this variation is unknown, but it shows that there must be continual mixing of material on the solar surface.

Although the spots show up as dark, they are not actually cold but are merely some 2 000 K cooler than their surroundings. Sunspots often occur in groups with two leading members. Near the spots there are frequently bright patches known as *faculae*, meaning 'little torches' (from the Latin). These faculae often endure much longer than the spots themselves. They are thought to be temporary mountains of hot gas rising above the solar surface. The cause of the spots still remains a mystery.

One of the principal discoveries about sunspots was made by

G. E. Hale at Mount Wilson in California. It was known that a magnetic field gives rise to a characteristic splitting of spectral lines. Hale found such an effect in sunspot spectra and concluded that sunspots have strong magnetic fields. Pairs of spots usually have opposite magnetic polarity. Records show that the numbers of spots visible each day oscillate more or less rhythmically in a cycle with a mean period of rather more than eleven years.

The Corona

At the time of a solar eclipse there comes into view one of the most beautiful sights in the skies, the pearly halo around the sun known as the corona. Shortly before the last war the French astronomer B. Lyot invented an instrument known as the corona-graph, by means of which the corona can be studied at any time and not only when there is a total solar eclipse; this instrument in effect produces an artificial eclipse. Although Lyot's corona-graph does not show all the details which are observable during a total eclipse, it can be used continually in order to study the sequence of changes of the corona. Moreover, with its aid cinematograph records can be obtained.

The structure of the corona is complex. It varies with the sun-spot cycle: at sunspot maximum it is compact, whereas at sun-spot minimum there are short streamers which, it has been suggested, may indicate the positions of the sun's magnetic poles. The sun's magnetism is, however, both weak and patchy and at the poles it is probably no stronger than the magnetic field of the earth. It is now thought that the study of the irregularities in the distribution of the sun's general magnetic field may eventually provide the clue to the formation of the long streamers which are sometimes seen emanating from the corona and also of the glowing clouds of vapour, known as prominences, which are often seen rising above the solar surface.

Superimposed on the spectrum of the corona are a number of bright emission lines which for long remained a puzzle to astro-physicists as they did not seem to be related to any known chemical element. Indeed at one time an artificial element, known as 'coronium', was invented to account for them. But as the science of chemistry and physics developed, the Periodic Table (Fig. 1) was gradually filled up and there seemed to be no place in it for this mysterious element. Astrophysicists came to the conclusion that these emission lines must be due to familiar substances distorted in some way by the curious physical conditions which it was presumed must prevail in the corona. Finally, in 1940, B. Edlén, a Swedish spectroscopist, put forward what is now thought to be the correct explanation, that these peculiar lines are due to highly ionized atoms which have lost a large number of their attendant electrons. By a brilliant combination of theory and experiment he showed that many of these lines are due to atoms of iron which have lost 9, 10, 12 or even 13 electrons. Neutral iron has 26 attendant electrons per atom, whereas the most prominent green line in the coronal spectrum is due to iron which has lost 50 per cent of its satellite electrons. Such a drastic stripping of electrons from atoms occurs at high temperatures, and the surprising conclusion of Edlén's investigation was that extraordinarily high temperatures must prevail in the corona. This result has been confirmed by the observation of the broadening of the spectral lines, which is again a typical consequence of high temperatures. For the observed stripping of 13 electrons from iron temperatures higher than 1 000 000 K are required.

We now recognize that the solar corona merges into the *zodiacal light*, a permanent feature of the evening and morning sky consisting of a faint cone of light extending along the zodiac. The zodiacal light is due to the scattering of sunlight by dust particles that form an interplanetary medium extending from near the sun to beyond the earth. The existence of dust particles near the sun has also been inferred from the discovery of excess infra-red radiation when the sun is observed at wavelengths of about 2×10^{-3} mm. This radiation is due to thermal emission from these particles.

Solar–Terrestrial Relationships

We not only receive heat and light from the sun but also material particles. This was first discovered by studying variations in the earth's magnetism known as magnetic storms. As long ago as

1859 Carrington at Greenwich found that certain bright flares on the sun were followed on the two following days by fluctuations in the strength of the earth's magnetic field. In the early nineteen-thirties Sydney Chapman and V. C. A. Ferraro put forward their well-known theory that in such a flare streams of protons and electrons, in more or less equal numbers and hence electrically neutral, are ejected from the sun. If a stream is directed towards the earth, it may impinge on the earth's atmosphere a day or more later. It has been suggested that these particles are deflected mainly towards the magnetic poles and generate powerful electric currents which tend to produce the magnetic storms mentioned earlier.

It has long been suspected that there is a relation not only between these terrestrial effects and solar flares, but also with sunspots. If such a relation exists, then it must be complex and somewhat elusive, because the presence of a large spot is not necessarily an indication that there will be a violent magnetic storm, and conversely such a storm can occur when there are no large spots on the sun.

The Solar Wind

One of the most exciting developments in astrophysics in recent years has been the discovery of a continual 'solar wind'. Clues to its existence, which had been predicted by Biermann, had been known for some time. Geomagnetic activity never ceases completely and there were other indications of solar control of an intervening plasma pervaded by a magnetic field. In 1957, Sydney Chapman pointed out that electrons in a hot corona would give so large a thermal conductivity that, if the temperature near the sun were about 1 000 000 K, the corona would extend past the earth with a temperature of 100 000 K. Chapman was still thinking of a static corona, but the following year E. N. Parker showed that there was no way of supplying the pressure required to maintain it. He concluded that a steady wind must blow from the sun dispersing the upper layers of the corona into outer space. He calculated that the outflow would be slow near the sun but that at a distance of a few times the sun's radius it would become supersonic, and at large distances it would tend to a limiting value of about 400 kilometres a second.

Parker's theoretical prediction was confirmed by observations by space probes, notably Mariner II in 1962. These and later observations revealed that a wind blows continually from the sun with a velocity varying between 250 and 800 kilometres a second, the motion being very nearly radial. The number of electrons or positive ions near the earth ranges from about 3 to 10 in each cubic centimetre, with occasional regions of somewhat higher density. Although a small magnetic field is present, there are indications that the solar wind does not originate predominantly in sunspot regions, where strong magnetic fields of opposite sign occur close together, but in extended magnetic regions where the field, although weaker, is directed outward or inward throughout. It appears that the wind can escape only from regions where the density of magnetic energy is less than that due to thermal motions, and it has been deduced that it is only at a height of about two-thirds of the sun's radius that the wind can break loose from its magnetic bonds. Properties of the solar wind have actually been used to predict with a fair degree of success the form of the corona to be observed at a particular eclipse.

Scientists think that convection is the principal means of energy transport just below the sun's surface and believe it to be connected with the origin of the solar wind. This conclusion is indicated by the cellular appearance of the solar surface which is thought to be due to the presence of both rising and falling quantities of matter. The surface layers look as if they were boiling, and the solar wind may be likened to a kind of evaporation from them.

The solar wind is thought to account for the girdle of rapidly moving particles surrounding the earth known as the Van Allen radiation belt, after the American physicist mainly responsible for the apparatus carried in an American artificial satellite and in the lunar probe Pioneer III which first revealed its existence. This belt, consisting mainly of protons and electrons, extends from a few hundred miles above the earth's surface to over thirty thousand miles. It is at its thickest and highest over the equator.

The charged particles in it are thought to be trapped by the earth's magnetic field and follow a corkscrew path forwards and backwards along the lines of force, possibly for several weeks. When a powerful outburst from the sun distorts the earth's magnetic field the particles with the most energy penetrate the earth's atmosphere and on colliding with atoms of oxygen and nitrogen at heights of 60 to 100 miles up, and occasionally higher, produce in extreme northern and southern latitudes the spectacular red, green and blue patterns of the Aurora Borealis.

The Sun's Gravitational Field

Since the Copernican theory came to be generally accepted, the other important invariable feature of the sun, its constancy of apparent motion, has been regarded as due to the constancy of the earth's motion about the sun under the influence of the sun's gravitational field. Given the earth's mean distance from the sun, the strength of the sun's gravitational field determines the length of the year.

The sun's superficial gravity is nearly 28 times that of the earth; so that a quartern loaf, if one could imagine it transported to the sun's surface without being affected by the heat, would weigh nearly a hundredweight, whereas if it were carried to the moon it would weigh less than three-quarters of a pound. The primary effect of the sun's gravitational field is to hold the solar system together. This was first made clear by Newton who followed Copernicus in regarding the sun as the central body. Decisive observational support for the Copernican theory was obtained by Galileo in 1610 when he first turned his telescope to the observation of the planets.

Venus, Jupiter's Satellites and Mars

The two planets which Galileo found to be most useful for confirming the Copernican theory were Venus and Jupiter. According to the Ptolemaic theory, Venus moved about the earth in such a way that always less than half of its surface appeared to be illuminated. The phases of Venus cannot be observed by the naked eye, but with his telescope Galileo discovered that, although Venus has phases, it is sometimes gibbous in appearance. This was a strong argument against the Ptolemaic theory.

Of even greater importance was Galileo's discovery of the four major satellites of Jupiter. One of the most powerful arguments that had been brought to bear against the Copernican theory was that, if all the other bodies in the solar system were to move around the sun, why should the moon be the only exception and move around the earth? Why did the other planets not have moons, too? Galileo wrote of his discovery, 'Jupiter removes the apparent anomaly of the Copernican system, that the earth was the only planet with a moon going round it.' Since 1892 eight more satellites of Jupiter have been discovered. They are very much smaller than the four Galilean ones, which are all of comparable size with our own moon. At times Jupiter's satellites can be seen casting shadows on Jupiter (see Plate XIV).

Although both Venus and Jupiter played a prominent part in helping to establish the Copernican theory, the planet that proved crucial for the development of the Newtonian theory was Mars. Prolonged study of the careful observations made (before the advent of the telescope) by Tycho Brahe (1546–1601) of the motion of Mars enabled Kepler to discover in 1605 that planets move in ellipses with the sun at a focus, a result which Newton showed some eighty years later could be explained if the sun attracted each planet gravitationally according to the inverse square of its distance. We now regard the ellipse as the form of the unperturbed path of each planet, neglecting the gravitational forces acting on it from the other planets.

Investigation of the Inner Planets by Radar and Rocket

We have already seen how radar waves reflected by Venus have been used to redetermine the linear scale of the solar system. Radar has also been used to study the rotation of that planet and has unexpectedly revealed that Venus spins in the opposite direction to nearly all the other planets and their satellites. Moreover, although it had been suspected for many years that the rotation of Venus was very slow, it was surprising to discover that the period of this rotation is about 250 days, since the orbital period is only 225 days. Consequently, although Venus is only slightly smaller and less massive than the earth its rate of rotation is far slower, being much less than that of any other body in the solar system, so far as we know. Why this should be so is a mystery.

Radar has also revealed a surprising result concerning the rotation of Mercury. From optical observations it had been inferred that, like our moon, Mercury rotates with the same period (88 days) as that of its motion around the sun. Radar observations have shown, however, that its rotational period is exactly *two thirds* of its orbital period. This is believed to be due to a peculiar resonance effect produced by the combined influence of the tidal action of the sun on Mercury and the unusually high eccentricity of Mercury's orbit, the distance of Mercury from the sun varying between 29 and 43 million miles.

From the earth, Mercury is one of the most difficult planets to view optically because of its closeness to the sun. It has no atmosphere and its surface appears to be similar to the moon's. Its radius is about two fifths that of the earth and its density is about 3·8 times that of water. In autumn 1973 it is hoped to launch a Mariner spacecraft equipped with cameras to take the first close-up pictures of Mercury, and also of Venus.

So far the only planet photographed by a passing spacecraft is Mars. Pictures televised in 1965 by the space-probe Mariner IV and in 1969 by Mariner VI and VII revealed numerous craters similar to those on the moon (see Plate XV). It was also found that, like the moon and Venus, Mars has no detectable magnetic field. Its radius is about half that of the earth and its mean density is about four times that of water. Its orbit is somewhat more eccentric than the earth's. The mean distance of Mars from the sun is about 142 million miles. Its year is equal to 687 terrestrial days, and its day is about half an hour longer than our day. Mars's equator is inclined to its orbital plane at a slightly greater angle than is the earth's and it therefore experiences seasonal effects, the most noticeable being the waxing and waning of the polar caps. In November 1971 Mariner IX began to orbit Mars (Mariner VIII never left the earth), the nearest point to the surface of the planet being only about 870 miles from it. The first pictures taken of the surface of Mars were badly blurred because of a severe dust storm. The maximum equatorial temperatures are about 30°C, and at night the temperature drops to about −70°C.

Generally, the Martian surface can be described as cold and arid, and if oxygen exists in the atmosphere it can only be as a trace constituent. There is a little water vapour and about thirty times as much carbon dioxide as in the earth's atmosphere. In 1969, over 200 pictures of Mars in close-up were taken by cameras on Mariner VI and Mariner VII. It was found that, although many of the craters are small, others are large and flat. These are thought to be very old and their existence implies that there have never been any oceans on Mars. Careful study of the south polar ice cap indicates that it is probably composed of frozen carbon dioxide a few feet thick.

Jupiter

Whereas the inner planets are not greatly different in size from the earth, which is the biggest, the outer planets (except Pluto) are considerably larger. The largest is Jupiter, its mass being 318 times that of the earth. Even so, it is still only one thousandth the mass of the sun. Besides being the most massive it is also the most rapidly rotating planet, its day being just under ten hours. Since its mean density is only about 1·3 times that of water (compared with the earth's 5·52), in all probability it consists mainly of hydrogen, most of it solid. It has been calculated that its volume (about thirteen hundred times that of the earth) is nearly the maximum possible for a body made of cold hydrogen, since both a less and a more massive body of this material would be smaller, the latter because the greater internal pressure would tend to strip the electrons from the atoms and consequently compress the body.

Jupiter is remarkable for its large red spot which varies in colour and size from time to time (see Plate XIV). Its maximum size is about 30 000 miles across. No explanation of its nature has yet become generally accepted. In 1955 Jupiter attracted much attention when it was found to be an intermittent source of radio noise. This was unexpected because of the low surface temperature of the planet. However, it is now believed that this radiation does not emanate from the main body of the planet but from a kind of Van Allen belt of charged particles circulating in a strong magnetic field associated with the planet. Jupiter is the only planet apart from the earth that is known to have a magnetic field.

Saturn

Jupiter with its numerous moons forms a solar system in miniature. So, too, does Saturn with its satellites, but its ring system is unique. The first person to see the rings was Galileo, although his telescope was not adequate to reveal their shape. It was Huygens who first succeeded, in 1655, in perceiving the ring form. Twenty years later Cassini resolved it into two concentric rings with a narrow dark division between them.

The rings are remarkably flat and thin and when presented to us edge-on we do not see them at all. They are not continuous solid or liquid structures, but consist of a swarm of discrete particles of matter each describing its own orbit about Saturn. It was shown by Clerk Maxwell in 1857 that continuous solid or liquid rings would be unstable and would break up under the tidal influence of Saturn. Only a swarm of separate bodies moving in nearly circular orbits in one plane could form a stable system. There is much observational evidence in support of this theory. The rings have been found to be excellent reflectors of light, and it is known that matter generally reflects better when pulverized. The constituent particles are probably quite small, like ordinary dust, but not so small as to be forced away from Saturn by pressure of the light which Saturn reflects from the sun. The American astronomer Kuiper has suggested that, as the particles have good reflecting surfaces, they are either composed of ice or else are covered with frost.

Saturn has ten satellites of which one, Titan, is remarkable in being the only satellite in the solar system with a detectable atmosphere. Kuiper believes that the atmosphere of Saturn originally extended beyond Titan which, owing to its comparatively large mass (nearly twice that of the moon), succeeded in retaining some of this atmosphere, and that the smaller satellites and the ring system were formed within this atmosphere as it cooled. The ring system may have been due to the disruption of a satellite which came too near to Saturn and was broken up by tide-raising forces into a number of small particles which finally settled down into a ring under gravitational action. Three rings have long been known, two bright ones with a narrow division between them and a fainter ring inside. Recently, a very faint fourth ring was discovered on photographs taken at the Pic du Midi observatory in the Pyrenees.

Saturn is about 95 times as massive as the earth. Its diameter is nearly 72 000 miles, and so its mean density is only 0·71 times that of water – the lowest of any planet in the solar system. Its internal structure is thought to be very similar to that of Jupiter, and it probably consists largely of hydrogen. It rotates nearly as rapidly as Jupiter and is the most oblate of all the planets.

The Discovery of Uranus and the Minor Planets

Until 1781 Saturn was the outermost planet known, but in that year William Herschel observed an object with a well-defined disc that at first he thought was a comet but when its orbit was computed it was found to be nearly circular, with a radius nineteen times that of the earth's orbit. It was therefore not a comet but a planet. This was the first discovery of a planet since prehistoric times. Herschel suggested that it should be named after the reigning king of England, George III, but eventually it was decided to call it Uranus. Its mass is about 14·5 times that of the earth and its mean density about 1·5 that of water. Its interior structure is probably similar to that of Jupiter and Saturn, but with a somewhat smaller proportion of hydrogen. It is unusual in that its axis of rotation lies nearly in the plane of its orbit. It has five satellites, two of which were discovered by Herschel, and one by Kuiper as recently as 1948.

The discovery of Uranus made the number of planets seven, and the philosopher Hegel endeavoured to prove that there could only be this number. But on the first night of the nineteenth century, the Italian astronomer Piazzi at Palermo discovered an eighth. He named it Ceres, after the tutelary deity of Sicily. It was a very faint object and soon after its discovery it was lost to view when it was overtaken by the sun. Piazzi tried to compute its orbit, but when he sought it again at the computed place, after the period of its invisibility, he could not find it. At this point, the great mathematician Gauss turned his attention to the problem and from Piazzi's observations determined the orbit by a new mathematical technique, with the result that by the end of the year the planet was recovered. Its orbital period is about four years, intermediate between that of Mars and Jupiter. Gauss's method soon found further application. In 1802, Olbers discovered a second small planet, named Pallas, having an orbit of nearly the same size as that of Ceres. In 1804, a third such planet, Juno, was discovered and in 1807 a fourth, Vesta. In this way the gap between the orbits of Mars and Jupiter, in which Kepler had suggested there might be an undiscovered planet, began to be filled – not by one planet but by a host of minor planets. Whereas the first four discovered were bodies of between 500 and 800 kilometres in diameter, many of the two thousand now known are less than 50 kilometres in diameter.

The minor planets all revolve about the sun in the same direction as the principal planets (from west to east) and most of them have orbits that lie nearly in the plane of the earth's orbit. An interesting feature in the distribution of these orbits is the existence of several clear regions known as Kirkwood gaps, after Daniel Kirkwood who in 1866 showed that they are due to the perturbational effects of the planet Jupiter. The situation is similar to that of the divisions in the rings of Saturn. The gaps occur where the orbital period would be a simple fraction, e.g. one half, two fifths, etc., of Jupiter's orbital period. On the other hand, there are no Kirkwood gaps when the orbital period is close to that of Jupiter. Instead, when the period is equal to that of Jupiter there is an accumulation of minor planets in place of a gap. This has been explained on the basis of a particular solution of the three-body problem in Newtonian gravitational theory, found by the French mathematician Lagrange two hundred years ago. Lagrange discovered that, if two of the bodies are each of small mass compared with that of the third and all three are located at the vertices of an equilateral triangle, then they can revolve in these same relative positions for an indefinite time about their common centre of gravity. Since the masses of the minor planets are very small, and even Jupiter's mass is only about one thousandth that of the sun, any minor planets located at the same distance from Jupiter and from the sun as Jupiter is from the sun will remain in the same relative position until perturbations by other planets disturb them. Two such groups of minor planets, known as the Trojans, have been found, one to the east and one to the west of Jupiter, members of the former being named after Greek Homeric heroes and those to the west after Trojan.

The origin of the minor planets presents an intriguing problem. Their combined mass may well be less than a thousandth that of the earth. It is therefore unlikely that they have been generated by the break-up of a *planet*. Moreover, if they had been so formed, one might expect their orbits to show evidence of this – for example, by intersecting at the same point. Nevertheless, the absence of any such evidence could be due to perturbations since the time of break-up. It has been found that many of the minor planets fall into families of similar orbital characteristics. In 1950 Dirk Brouwer found that some of these families can be divided into two sub-groups, which suggests that originally there was a collision between two smallish bodies. It can easily be shown mathematically that subsequent collisions and fragmentation would tend to occur with increasing frequency. The German astronomer Unsöld has referred to this phenomenon as a 'solar-system gravel mill'. The ultimate future result would seem to be the production of a vast amount of fine dust.

Neptune, 'Vulcan' and Pluto

The most remarkable discovery of a planet in modern times was that of Neptune in 1846. For some years it had been known that certain perturbations in the orbit of Uranus could not be accounted for by the action of known planets. The mathematical problem of explaining these by the gravitational attraction of an unknown trans-Uranian planet was tackled by Adams in this country and Leverrier in France. The story of the discovery of a planet near the predicted place and the resulting uproar as to whom priority of discovery should be assigned is well known. At the time the mathematical prediction of Neptune on the basis of Newton's theory was regarded as one of the greatest achievements of the human mind. Nowadays, we assess this much more soberly and attach greater importance to another, and apparently far less successful, calculation made by Leverrier some years later on the motion of the planet Mercury.

The theory of Mercury's motion had long given trouble so that, as the historian of nineteenth-century astronomy Agnes Clerke said, 'The planet seemed to exist for no other purpose than to throw discredit on astronomers!' Even to Leverrier's powers of analysis it long proved recalcitrant, but in 1859 he announced that a body of about the size of Mercury revolving at somewhat less than half its distance from the sun (and therefore very difficult to detect) would produce the required effect of advancing the perihelion of Mercury's orbit (that is the point in this orbit nearest the sun – chosen as a conveniently identifiable point) by the 38 seconds of arc in a century that could not otherwise be explained. From time to time, claims were made that this planet, to which Leverrier gave the name 'Vulcan', had been seen. All proved to be illusory, but the reality of the perihelion effect could not be denied. In 1884, the American astronomer Simon Newcomb corrected Leverrier's 38 seconds to 43 seconds, but thirty years were to pass before Einstein unexpectedly found that the phenomenon was completely accounted for by his general theory of relativity. This was indeed the first empirical confirmation of his theory of gravitation (see Chapter 6).

With the discovery of Neptune all but about 2 per cent of the discrepancy between the predicted and observed positions of Uranus were accounted for. Nevertheless, several investigators tried to account for this residual effect by the gravitational influence of a trans-Neptunian planet. In 1930 this was discovered by Clyde Tombaugh within 6 degrees of the position predicted by Percival Lowell. The new planet was named Pluto. It was found to have an orbit which is highly inclined to that of most other planets and very eccentric. Although Pluto's mean distance from the sun is nearly 40 times the earth's, whereas Neptune's is about 30 times, part of its orbit is closer to the sun than the orbit of Neptune. Its 'year' is equal to nearly 250 terrestrial years. Its mass is similar to that of the earth and some astronomers have conjectured on this basis and from the fact that its orbit intersects Neptune's that it may be an escaped satellite of Neptune, whose mass is about 17 times that of the earth. (Neptune is known to have two satellites, one of which moves in a highly eccentric orbit.) Ironically, Pluto's mass is too small to produce the perturbations on which Lowell's prediction was based. As the American astronomer George Abell has said, 'The fact that Pluto *was* discovered, with so nearly the orbital elements that Lowell predicted, may be one of the most startling coincidences in the history of astronomy. Be that as it may, it was Lowell's faith and enthusiasm that led to our knowledge of Pluto.' As for the unexplained small discrepancies in the motion of Uranus, it is now thought that they may have been due to slight errors in the calculations of the perturbing effects of the other planets.

The Moon

The earth's nearest celestial neighbour, the moon, accompanies the earth in its annual revolution around the sun. The moon's period of revolution about the earth with respect to the stellar background is approximately 27 days 7 hours 43 minutes, but during this time the earth-moon system has moved through about one thirteenth of its revolution around the sun. Consequently, to complete a revolution about the earth with respect to the sun the moon requires, on the average, approximately 29 days 12 hours 44 minutes. The former period is known as the *sidereal month* and the latter as the *synodic month*. A very accurate determination of the moon's distance was made by the radar technique in 1957 and the mean distance thus obtained from the centre of the earth to the centre of the moon – 238 857 miles – is believed to be correct to within a mile. An even more accurate determination has since been made with a laser beam reflected from a retro-reflector placed on the moon in 1969. The moon's mass is 8.1×10^{19} tons, which is one part in 81.3 that of the earth.

Its mean density is 3·34 that of water, or just over three fifths that of the earth. Not only is there no trace of any appreciable atmosphere but there is also no water on the moon. Since water boils at a lower temperature the lower the atmospheric pressure, it follows that in the absence of the latter it would immediately evaporate into a gas. Consequently, any liquid water there might once have been on the moon would have evaporated long ago and been dispersed into space with the moon's primeval atmosphere. After Frank Press stated categorically, in the Harold Jeffreys Lecture delivered at a meeting of the Royal Astronomical Society on 12 March 1971, 'There is absolutely no evidence for any water in the lunar rocks,' John Freeman announced that the most probable explanation of the data obtained concerning a cloud of gas observed on the moon on 7 March 1971 was that it contained water molecules. The cloud seems to have been produced by a series of seismic events, and it is possible that ammonia, neon or other rare gases, rather than water vapour, may have been responsible for the observed phenomena.

The moon rotates on its axis with the same period as it revolves around the earth. This is believed to be a long-term consequence of the earth's tidal force on the moon. As a result we can see from earth only one half of the moon's surface. On this side some thirty thousand craters have been observed from earth, the largest being nearly 150 miles in diameter. Still larger are the maria, great plains that appear darker than the surrounding regions and are believed to have originated from great lava flows. Two hypotheses have been suggested to account for the craters: either they have a volcanic origin or they are due to explosions following the impact of fast moving meteorites.

The first pictures of the far side of the moon were taken from the Russian rocket vehicle Lunik III in October 1959 when it successfully passed behind the moon. Broadly speaking, the two sides appear to be similar, except for the relatively smaller area covered by maria on the far side.

One of the unexpected discoveries resulting from orbital flights around the moon in the late nineteen-sixties was the discovery of 'mascons', regions in the circular maria with increased gravitational attractive force. Some selenologists believe that they are caused by enormous iron or nickel meteorites up to fifty miles in diameter and buried at about this depth below the lunar surface. Others argue that, after the filling of the maria basins with lava, layers of high density were produced by phase changes or the settling of heavy oxides. Another possible explanation is that the presence of mascons is an indication that the moon is a comparatively loose agglomeration of interplanetary debris.

Orbital flights around the moon have, however, already resolved one long-standing problem. For half a century the study of the twisting of the lunar orbit (precession of the nodes) seemed to point to a value for the moon's moment of inertia that indicated it to be hollow! It now appears, however, from the new value obtained for the moment of inertia that the distribution of matter in the moon is uniform, unlike that in the earth which is much denser in the central core than in the outer mantle.

A new era in man's exploration of the universe dawned on 21 June 1969 when the Americans succeeded in their Apollo XI mission and successfully landed two astronauts on the moon. By the end of 1972 four further landings had been made and on each occasion samples of lunar rock were collected and brought back to earth for detailed laboratory examination. As a result it has now definitely been established that the maria are filled with basalts similar to the great lava flows in Iceland, Northern Ireland and elsewhere on earth. Petrologists have found that these basaltic rocks were formed by a process of fractional crystallization similar to that in terrestrial lava which we know is due to partial melting below the earth's surface due to internal heat sources. Chemists have studied the radioactive elements present in the lunar samples and have found that the lavas of the Sea of Tranquillity cooled 3 700 million years ago, whereas those of the Sea of Storms are nearly a thousand million years older. The age of the moon is believed to be the same as that of the earth – about 4 500 million years. However, whereas atmosphere and ocean cause relatively rapid changes to the earth's surface, the lunar surface is eroded slowly by bombardment by meteorites, the solar wind and cosmic rays. Also continental drift has radically changed the disposition of the great land masses on the earth's surface within the last two hundred million years. On the moon, owing to the apparently complete absence of water in the interior (as well as on the surface) and the likelihood that the temperature even at the centre is less than 1 000°C, the conditions that give rise to continental drift on earth do not occur and it would seem that the surface of the moon has not changed significantly in the last 3 000 million years. This conclusion is inferred from the absence of any detectable magnetic field on the moon. Planetary magnetic fields, such as those of the earth and Jupiter, are believed to arise from motions in their liquid metallic cores where electric currents are generated as in a dynamo. It is thought that the heat energy which keeps the terrestrial dynamo going may be supplied by the radioactivity inside the earth. The lava flows show that the moon's interior was once hot but all the indications are that that era ended several thousand million years ago.

Support for the view that, whereas the earth shows every sign still of significant internal mobility (which we now believe is responsible for earthquakes, mountain building and continental drift), the moon is inert, has also come from the study by physicists of the tracks made by atomic particles in the minerals brought back by lunar astronauts. They conclude that the lunar 'soil' has lain undisturbed for millions of years.

The fact that the age assigned to the oldest rock from the Sea of Storms is 4 600 million years indicates that some parts of the lunar surface have remained essentially unchanged since the moon was originally formed. Consequently, the lunar surface, unlike that of the ever-changing earth, provides us with a hitherto unavailable record of the early history of the solar system and that is why it is important that we should study it in the closest possible detail.

Comets, Meteors and Meteorites

There are many other constituents of the solar system besides planets, satellites and minor planets, notably comets, meteors and meteorites. Together with interplanetary dust they all bear witness to the existence of a certain amount of diffuse material within the system.

Nearly all comets move in highly elongated elliptical orbits around the sun and consequently are not, as was at one time suggested, intruders from outer space but are true members of the solar system. The most famous comet, named after the astronomer Halley, has an orbital period of about 76 years and is due to reappear in 1986. The comet of the shortest known period is Encke's, 3·3 years. There are two comets known to have nearly circular orbits which in consequence have been under more or less continual observation from the earth. One lies between Mars and Jupiter, and the other between Jupiter and Saturn. The latter, known as Schwassmann-Wachmann is normally rather faint but from time to time it suddenly flares up to more than one hundred times its normal brightness. This phenomenon is thought to be associated with violent emissions of charged particles from the sun.

Despite their awe-inspiring appearance when near the sun, comets are very insubstantial bodies compared with planets. Although no one has yet been able to determine the mass of a single comet, it has been estimated that it might need nearly a billion comets to yield a mass equal to that of the earth. One of the most remarkable comets was that of 1882, which was plainly visible in full daylight. Since it was not possible to see it when it was in front of the sun, its nucleus must have been *less* than 50 miles in diameter. All comets are surrounded by a mass of gas, the *coma*, which consists of molecules and atoms liberated from the nucleus by the sun's heat. The American astronomer F. L. Whipple has likened a comet to a 'dirty iceberg', which is a mixture of ices of water, ammonia, methane, etc. with grains of metals (e.g. iron) and stones. As a comet approaches the sun the lighter molecules evaporate. Whipple has calculated that on each passage around the sun a comet loses about one part in two hundred of its mass in this way. For the average comet he has estimated its nucleus to be a sphere of only about a kilometre in radius.

Many comets develop tails as they near the sun. Generally these tails point away from the sun, some attaining lengths of more than 100 million miles. For many years the force that drove the particles in cometary tails away from the sun was thought to be radiation pressure, but it was impossible to account in this way for the larger accelerations which sometimes exceed the force of gravity more than a hundred-fold. It is now thought that these large accelerations must be due to the effect of the solar wind.

The most popular explanation for the origin of comets is that suggested by the Dutch astronomer J. H. Oort. He believes that there is a vast cloud of comets, running into many thousands of millions, most of which move in orbits between 50 000 and 150 000 times the distance of the earth from the sun. At more remote distances they would soon be lost to the solar system. On the other hand, those that are retarded in their motions by the gravitational influence of the stars fall in towards the sun. Some of these move in orbits that bring them close to the sun. Eventually, they are either deflected by the planets, for example Jupiter, and lost to outer space or else they continue to move in orbits bringing them near the sun and survive only a few million years before completely evaporating.

The probability of our colliding with the nucleus of a comet is small, but the earth probably passed through the tail of Halley's comet in 1910. Although there was no noticeable effect, it was afterwards reported that someone had made a fortune by selling 'comet pills'!

In the middle of August each year the earth passes through a swarm of particles that approach from the direction of the constellation Perseus. In November there is a similar shower of meteors known as the Leonids. Occasionally, the profusion of meteors observed in such a shower, as for example on the night of 13 November 1833, is phenomenal. On that occasion some stations reported that there were as many as 200 000 an hour. From an analysis of orbits of some showers we know that they originated in comets that have disintegrated. On the other hand, twice a year the earth passes through a swarm of particles moving in the orbit of Halley's comet.

Following the nomenclature recommended by the International Astronomical Union in 1961, the term 'meteor' should be restricted to the luminous phenomenon ('shooting star') caused when a particle vaporizes in the earth's atmosphere and when it is in space it should instead be called a '*meteoroid*'. If it is sufficiently massive to survive passage through the earth's atmosphere and land on the ground, it is called a '*meteorite*'. The mass of a meteoroid producing a typically bright meteor is about a quarter of a gramme (one hundredth of an ounce), but many meteors are produced by particles of only a few milligrammes. Although meteorites tend to be very much bigger, many extremely small micrometeorites land on earth each day. They are only a few thousands of a millimetre in diameter and are slowed by the air before they ever have a chance to get hot. The total fall of micrometeorites has been estimated to be of the order of 2 million tons a year over the whole globe.

Meteorites are of two kinds, one consisting primarily of iron and nickel and the other of rock. The former are particularly interesting because they have a peculiar crystalline structure which becomes visible as soon as a polished cross-section is etched by an acid. This Widmanstätten structure has been shown to be due to solidification following extremely slow cooling from a relatively high temperature, from which we can conclude that iron meteorites are fragments of a much larger parent body. Consequently, whereas meteors are associated with comets, meteorites are thought to result from the disintegration of minor planets. Natural radioactivity provides a means of determining the ages of meteorites just as it provides a means for determining the ages of rocks in the earth's crust. The relative concentrations in a meteorite of, say, uranium, helium and radiogenic lead indicate how long the radioactive decay process has been going on and hence the age of the meteorite. Some meteorites give ages of less than a million years, but the maximum age of meteorites comes out to be about 4·5 thousand million years, which is believed to be the age of the earth and of the solar system.

The largest meteorite ever found (near Grootfontein in South Africa) weighed about 50 tons, but there is evidence in various parts of the world that much larger meteorites have fallen, the best known being that which produced the Barringer crater in a plateau of stratified limestone and sandstone in Arizona perhaps fifty thousand years ago. This crater is about four fifths of a mile across and nearly six hundred feet deep. Estimates of the mass of the object that made it vary from several thousand to several million tons.

The two most spectacular falls of meteorites this century occurred in Siberia, on 30 June 1908 and on 12 February 1947. The former was in a marshy region in central Siberia, in the basin of the Tunguska river some 500 miles north of Lake Baikal, and the latter near Vladivostok. In both cases trees were felled radially, in the former over an area of more than 20 miles' radius, and in the latter in circles around more than a hundred craters, the entire region covering two square miles. Although more than five tons of iron fragments have been recovered from the 1947 Siberian meteorite, no pieces have been found of the 1908 one, although its original mass before it entered the earth's atmosphere has been estimated at not less than a hundred tons. Its fall had an effect comparable with that of a moderately strong earthquake. The fact that, despite the evidence of great damage to the surrounding forest, no trace of a crater has been found has led to much speculation about the nature of the Tunguska meteorite. In view of the evidence of great heat generated by the impact, it is thought the meteorite was moving in a retrograde orbit associated with a high relative speed. Such an orbit would not be typical of meteorites derived from the disintegration of minor planets. One suggestion was that this meteorite was in fact a small comet. Evidence for this hypothesis was the unusual luminescence of the night sky immediately after the fall over Siberia, Russia and Western Europe. Nevertheless, it is difficult to understand why the comet had not been previously observed, unless it approached the earth from a direction very close to that of the sun. At nearly 20 kilometres from the epicentre, trees were subjected to a thermal flash and started to burn, and from this it has been calculated that the total thermal energy of the explosion was not less than 10^{16} joules. This result is in good agreement with other estimates of the energy involved. It has proved difficult to construct a convincing model for either a chemical or a nuclear reaction to explain all the phenomena associated with the Tunguska meteorite. Consequently, more than once the suggestion has been made that the intruding body was composed of antimatter (see Chapter 1), the magnitude of the explosion being due

201

to automatic 'annihilation' of the meteorite when it interacted with the gases in the atmosphere. A very careful examination of this suggestion was made in 1965 by the radiocarbon experts W. T. Libby, C. Cowan and C. R. Athuri on the basis of the number of neutrons that could have been released into the atmosphere and the consequent effect on radiocarbon activity. Comparison with the observed radiocarbon content of tree-rings formed around 1908 led them to conclude that at most only one seventh of the energy of the Tunguska meteorite could have been due to anti-matter.

From time to time meteorites have been examined with a view to discovering traces of extra-terrestrial life. In 1969 a meteorite that fell in Australia was found to contain traces of amino acids, the basis of proteins. As the age of this meteorite, known as the Murchison meteorite, was about 4 500 million years – the age of the solar system – this discovery was hailed as support for the chemical theory of biological evolution, according to which various electrical discharges cause complex chemical molecules to evolve into biological molecules.

Planetary Atmospheres and Biological Evolution

The theory of the existence and chemical composition of the atmospheres of celestial bodies is based on the concept of velocity of escape. The velocity of escape at the surface of a body is that velocity with which a particle must be projected if it is to leave the body and never return to it despite the influence of gravity. At the sun's surface this velocity is about 55 times as great as the corresponding velocity on the surface of the earth. If a rocket is to leave the earth for outer space, it must be projected with a speed in excess of 11·3 kilometres, or about 7 miles, a second. To leave the sun it would have to be ejected with a speed greater than 622 kilometres a second, whereas on the moon it need only be projected with a speed greater than 2·4 kilometres a second.

For a particular gas to be retained for many millions of years in the atmosphere of a planet, or other body, it has been calculated that the *average* velocity of its constituent molecules must not exceed about *one fifth* of the velocity of escape. Otherwise, as a result of continual collisions, almost all the molecules will in time acquire the velocity of escape when near the boundary of the atmosphere and be dissipated into outer space. The average velocity of the molecules varies from gas to gas and also depends upon temperature. As already mentioned, the higher the temperature, the greater the average velocity.

Therefore, a planet near the sun, like Mercury, will have a worse chance than a planet farther away of retaining any particular gas, because Mercury is exposed to a much higher temperature from the sun's radiation. Also Mercury is a comparatively small planet and consequently has a weaker gravitational field. Calculations on these lines suggest that neither Mercury nor the moon should have any atmosphere, results that are confirmed by observation. On the other hand, Venus, Mars, Jupiter, Saturn, Uranus and Neptune all have atmospheres. The only satellite with an atmosphere is Titan, as already mentioned.

The surface temperatures of Jupiter and Saturn are much lower than those prevailing on the earth, an average of $-120°C$ on Jupiter and $-140°C$ on Saturn, those on Uranus and Neptune being even lower. The molecular velocities are correspondingly lower and, moreover, the gravitational fields are stronger. Hence, unpleasant gases like methane and ammonia are retained in the atmospheres of these bodies. It is therefore most unlikely that there are any living organisms on them (or on Titan).

The planet which is most nearly equal to the earth in size and mass is Venus, but unlike the earth it is perpetually covered with thick clouds. They consist largely of carbon dioxide and this is believed to be due not to the presence of plant life but to the high surface temperature which encourages chemical reactions between minerals containing carbon and oxygen. Conditions on the surface are governed by the greenhouse effect: a considerable part of the solar heat which falls on Venus is trapped beneath the clouds. For many years it was thought that either the planet was a desert body or that there was much water on the surface nearly at boiling point but kept at a fairly constant temperature by the cloud blanket. The issue was resolved in 1967 when the Russians succeeded in landing a space probe on Venus. They found that the surface is at a temperature of 430°C and sustains an atmospheric pressure nearly one hundred times as great as that on the earth. Venus is therefore presumably a lifeless desert.

The main reason why conditions on Venus, a planet only slightly smaller than the earth, should be so different is because it is nearer the sun. Its orbit is very nearly a circle of radius 67 million miles, compared with the earth's mean distance from the sun of 93 million miles. Water that might otherwise have formed oceans on Venus is still trapped in the atmosphere in the form of vapour. It has been calculated that, if the earth had been on the average only five million miles nearer the sun, its temperature would have been too high for water vapour to have condensed to form the oceans, and the earth would have been like Venus.

The most favourable extra-terrestrial place in the solar system to seek for atmospheric conditions which might be consistent with the existence of life is Mars. One of the most convincing arguments for assuming that Mars has an atmosphere is its appearance when photographed in the red and in the blue. Mars appears larger in blue light than in red, the reason being that planetary atmospheres, like that of the earth, tend to scatter sunlight in the blue. Comparison of the photographic plates leads to the conclusion that the depth of the Martian atmosphere must be about sixty miles. Although there is some uncertainty concerning its composition, the density at its base is believed to be comparable with that at the top of Mount Everest, where nothing can live under natural conditions. Hence, it seems unlikely that there is much life on Mars, except possibly some very primitive vegetation. Some years ago when studying the spectra of the planets with the aid of photo-electric cells of high sensitivity in the infra-red, Kuiper found that the greenish areas on Mars show no trace of the infra-red spectrum of chlorophyll, the green colouring matter in terrestrial vegetation.

The problem of extra-terrestrial life has attracted much attention in recent years, particularly in Russia. Biologists have carried out experiments in Pamir where the climate corresponds to that of the middle latitudes on Mars. In the valleys the mean annual temperature is below freezing point, but the air is very dry. The conditions prevailing there change plants which normally have considerable transpiration into forms with hardly any. Consequently, the dry and cold climate of Mars cannot be regarded as excluding the possibility of some form of vegetation. Nor does the absence of chlorophyll preclude this, for it has been discovered in Siberia that one variety of pine and other species of trees absorb chlorophyll only when the temperature is above freezing point. When it drops below that, no trace of chlorophyll can be detected. The possibility that some form of vegetation exists on Mars has been strengthened by some remarkable spectroscopic evidence obtained by Sinton, using the 200-inch telescope on Mount Palomar. Sinton discovered that certain infra-red radiation is absorbed in the dark areas of the surface of Mars but not in the bright ones. Two of the three absorption bands were known to be shown by molecules in which carbon and hydrogen atoms are linked. The third has since been found to be shown by cellulose, which is present in plants. Although not conclusive, Sinton's discovery is the most important evidence so far for the existence of plant life on Mars. The absence of oxygen, however, makes animal life, as we know it on the earth, impossible. Moreover, there can be little ozone, and this means that probably there is a substantial penetration of ultra-violet light to the surface. Although this must pose an additional hazard for life, some micro-organisms certainly could live on Mars.

The evolution of terrestrial life has been closely dependent on the history of the concentration of oxygen in the atmosphere. All lines of evidence suggest that the present atmosphere is not primordial but has been developed from secondary sources self-contained in the earth. Physical chemists (H. C. Urey and others) have argued that the compounds comprising the outer layers of the earth were never molten. Consequently, they retained large quantities of chemically bound gas. The earth's primitive atmosphere was generated from these by volcanic action. This atmosphere contained water vapour, carbon dioxide, etc. but no oxygen, since none is released directly from volanic effluents.

The oxygen now in the earth's atmosphere is due to photosynthesis. Indeed, the presence of oxygen at the stage when amino acids are formed in the presence of ultra-violet light would have tended to break them down again. On the other hand, photosynthesis can proceed in the absence of oxygen. In the earth's primitive atmosphere lethal ultra-violet solar radiation may have penetrated five to ten metres depth of water. But not enough light for photosynthesis could penetrate deeper. Considerations along these lines suggest that the earliest life consisted of green algae or their evolutionary precursors in protected shallow lakes or pools rather than in the oceans. Warm pools associated with volcanic hot springs rich in nutrient minerals may have been particularly suited for the primeval development of living organisms.

When the oxygen level rose to about one per cent of that in our present atmosphere it reached the 'Pasteur point' where organisms change from fermentation to respiration. At the same time the penetration of lethal ultra-violet solar radiation diminished to a few centimetres of water, and the oceans became more suited to sustain life. It we look into the geological and palaeontological record we find that the age corresponding to the onset of these conditions with the opportunities they provided for a proliferation of new organic forms was the Cambrian, beginning about 600 million years ago. Prior to the Cambrian age there is no evidence of any life beyond elementary algae, fungi and bacteria.

After the beginning of the Cambrian age, the complexity of life increased rapidly and more than a thousand new species appeared. With widespread photosynthesis by marine life the oxygen level rose, and when it had attained about ten per cent of its present concentration the dry land was at last shielded from dangerous ultra-violet radiation. The geological record shows no trace of advanced life ashore before the late Silurian age, about 420 million years ago. Then a number of different phyla of plants and animals suddenly appeared on dry land and by the early Devonian age some 30 million years later great forests sprang up and soon afterwards the first amphibian vertebrates appeared. We may therefore explain this stage of evolution by the oxygen in the atmosphere having attained one tenth of its present level. Rapid proliferation of life ashore then increased photosynthesis considerably.

This brief description of the possible interdependence of the evolution of terrestrial life and of the earth's palaeoatmosphere is based on the work of L. V. Berkner and L. C. Marshall, who suggest that on Mars, owing to the absence of oceans, the atmosphere may be somewhat similar to the pre-Cambrian atmosphere on earth. This view contrasts sharply with the idea of Percival Lowell earlier this century that current conditions on Mars correspond to a much later stage in biological evolution than we are now experiencing on earth and that the notorious 'canals' – the existence of which is now quite discredited – represent the heroic attempts of Martians to maintain civilization on 'a dying planet'!

The Origin of the Solar System

There is still no agreed theory on how the solar system originally came into being, and it may well be that none will ever be generally accepted. Some theories are based on the assumption that very peculiar conditions were required to originate such a system, but as these conditions must have existed a very long time ago there may be no comparable situation today by which these theories could be directly tested.

There is some indirect evidence for the hypothesis that there was much more diffuse interplanetary matter – gas and dust – scattered throughout the solar system in the distant past than there is today. Except for the extreme cases of Mercury and Pluto, the eccentricities of the orbits of the planets are so small that they differ very little from circles. It is known from theory that if at one time there existed a cloud of diffuse material throughout the system, then resistance to motion would have had the effect of progressively diminishing the eccentricities of the planetary orbits, which would gradually have become more and more circular. Since there is no reason why the primeval orbits should have been very nearly circular, it is thought that there must once have been much more diffuse interplanetary material than there is today.

More than a century and a half ago Laplace put forward his famous nebular hypothesis, according to which the solar system originated in a vast diffuse rotating mass. As this condensed under its own gravitational action, rings split off and condensed into planets. Although this hypothesis fell out of favour for a time, it has recently been revived in a modified form.

Meanwhile many other hypotheses have been put forward. According to one, the sun passed at some time very near another star which drew out material from it and this filament ultimately condensed into separate planets. The very close approach of two stars which were formerly far apart must, however, be a very rare event. Nevertheless, the main objection to this hypothesis is its failure to account for the peculiar distribution of rotational, or angular, momentum in the solar system, which is mainly carried by outer planets of far less mass than the sun.

Another hypothesis is based on the assumption that the sun was originally a member of a binary system. The other star, having suffered a direct collision with a third star, broke up and the planets were formed from the resulting fragments. However, in view of the high temperatures that might be expected to arise, it is more probable that the material would just stream out into space than condense into planets.

Two important recent theories, put forward by C. F. von Weizsaecker and H. C. Urey, respectively, are similar to Laplace's theory. According to von Weizsaecker, the sun passing through an interstellar cloud of diffuse material captured part of it. Numerous eddies formed in this cloud, which became very turbulent. From a detailed consideration of the vortices that might occur, von Weizsaecker has indicated how the planets and satellites could have been generated, but most investigators have been at a loss to understand how turbulence, which always signifies disorder, could have been the agent that created order in the solar system.

Whereas in von Weizsaecker's theory the solar system began when the sun was already a hot star, in Urey's the whole system began at a low temperature. The sun was created at the same time as the rest of the solar system from some kind of cloud in which condensations occurred, the largest giving rise to the sun and the others to the planets and satellites. The largest condensation, being so much more massive than the others, eventually became a star and raised the temperatures of the smaller bodies in the space surrounding it. However, their surface temperatures never rose to much above their present values. The conditions which prevail within this range of temperatures give rise to much more complex phenomena than those associated with the high temperatures of the sun and stars. The study of these phenomena is primarily the concern of the chemist, and Urey's approach to cosmogony was novel, being that of the chemist rather than that of the mathematicians and physicists who alone had studied the subject before.

Most astronomers today accept Urey's basic idea of a 'cold' origin of the solar system, for otherwise it is difficult to explain the internal constitution of the earth. Also, unless temperatures in the outer regions of the solar system have always been low, it is not easy to account for the presence of ammonia, methane, etc. in the major planets. Since these planets are believed to be more like the primeval gas cloud from which the solar system was formed than are the earth and other inner planets, it is particularly to be hoped that the Pioneer probes due to be sent to Jupiter in 1972 and 1973 will be successful in transmitting pictures back to us of that planet taken in close up. It will take just under two years to reach Jupiter on these flights. In 1977 and 1979 two further unmanned flights to the outer planets are planned by the U.S. National Aeronautic and Space Administration (N.A.S.A.). They will try to exploit rare configurations of these planets that will enable the spacecraft to be swung round by the gravitational attraction of Jupiter to Saturn and Pluto in the former case and to Uranus and Neptune in the latter.

Theories of the origin of the solar system fall into two classes, depending on whether they do, or do not, depend on peculiar initial conditions. In recent years there has been a growing tendency to regard the problem not in isolation, as used to be customary, but in relation to the general problem of star formation. For example, it is now thought that stars like the sun probably originate in groups of perhaps several hundred which are gradually dispersed owing to the shearing effect produced by the tidal action of the stellar system as a whole. Moreover, some theories invoke electromagnetic forces as a possible mechanism for transferring angular momentum from the sun to the planets.

But, although a condensation in a vast cloud of gas would, at some stage, be highly ionized and have a high electrical conductivity (so that it would be what is called a plasma), it is not yet possible to say whether this will enable us to account quantitatively for the origin of the solar system.

Throughout the universe there are millions of stars like the sun and many astronomers believe that there must be many other planetary systems. S. S. Kumar of the University of Virginia has, however, disputed this conclusion. He has drawn attention to the critical role played by Jupiter (the most massive of the planets, but only one thousandth of the sun's mass) in the stability of the solar system. According to his calculations, a slight change in the mass or orbit of Jupiter could upset the stability of the orbits of the inner planets, including the earth, and result in their collision with either the sun or Jupiter. If Jupiter were forty or fifty times more massive, the inner planets could survive for a total period not exceeding four hundred million years, less than a tenth of the currently accepted estimate of the age of the earth. Kumar concludes that the total number of planetary systems like our solar system must be small compared with the total number of stars.

Despite the spectacular advances in observational technique in the present century, however, no conclusive observational evidence of any other such system has yet been obtained. For all we know at present, the solar system may be unique. And it may well be that we shall never know whether there exists a world similar to ours elsewhere.

Great as are the distances between the sun and the planets, they are minute compared with the vast distances of interstellar space. Light from the sun takes about eight minutes to reach us on the earth. It can span the distance right out to Pluto on the apparent confines of the solar system in about five and a half hours, but light from the nearest known star, Proxima Centauri, takes more than four years to reach us.

CHAPTER 10

The Milky Way

Wright, Kant and Herschel

Until the latter part of the eighteenth century the principal object of astronomical investigation was the solar system. The stars were regarded as forming a mere framework of reference against which the motions of the planets and other bodies in the solar system could be studied and measured. Copernicus, Kepler and Galileo, by transferring the origin of reference from the earth to the sun, have been regarded as responsible for one of the greatest revolutions in the history of man's ideas concerning the universe and his relation to the universe. Nevertheless, despite its shattering impact on human thought, the Copernican revolution was essentially a rearrangement of the same pieces on the same chessboard, whereas the astronomical revolution of the second half of the eighteenth century substituted an entirely new chessboard for the old one and an entirely new set of pieces. The background now came into the foreground of attention and the solar system shrank into comparative insignificance.

This intellectual revolution was due primarily to the researches of one of the greatest observational astronomers of all time, William Herschel, whose pre-eminence has been recognized this century, by the laying of a memorial stone on the floor of Westminister Abbey in 1954. Greatly as he was esteemed by his contemporaries, the full scale of his achievement could not be properly assessed until our own day.

Herschel was a German from Hanover who became naturalized British. In his general conception of the role and nature of the Milky Way he had two immediate predecessors, Immanuel Kant of Königsberg, and Thomas Wright of the County of

Durham. It is due to Kant that Wright's name is not forgotten. In 1750 Wright published the one work for which he is famous, a short book entitled *An Original Theory, or New Hypothesis of the Universe*. In this he argued that the phenomenon of the Milky Way, or the galaxy as it is often called nowadays, is not due to any crowding together of stars but is purely an optical effect. Assuming the star system to which the sun belongs to be much less extended in one particular direction than in all directions perpendicular to it – so that in fact it is more like a bun than an orange – he showed that, if we are somewhere near the centre, then the stars will appear to be distributed with greatest density in a circular band running round the sky, just as the Milky Way does.

Kant never actually saw Wright's book, his knowledge of it being entirely due to an abstract in a Hamburg journal which happened to come his way. He seized hold of Wright's original idea and on it based his *General Natural History and Theory of the Heavens*, which was published anonymously in 1755. Wright's exposition was mainly geometrical, whereas Kant's was both dynamical and evolutionary, his object being to show how, on the basis of Newton's law of universal gravitation, the universe might have evolved into its present condition from an initial uniform distribution of particles.

The necessary observational support for the hypotheses of Wright and Kant was provided by William Herschel in the last decades of the eighteenth century and the first of the nineteenth. By his investigation of visual binaries, or double stars, he obtained the first definite evidence that Newtonian gravitation operated beyond the bounds of the solar system. With the aid of a more powerful reflecting telescope than any previously constructed he spent many years in his systematic sweeps of the heavens. Herschel's genius lay not only in his ability as a telescope constructor, but also as a visual observer. As he himself said, 'Seeing is an art.'

One of Herschel's main problems – indeed, his basic problem – was the determination of stellar parallax, i.e. the determination of the distances of the stars by measuring the angles of parallax which they subtend at the earth in the course of a year as the earth revolves around the sun. In order to study this problem Herschel found it necessary to investigate pairs of stars which appear to be close together in the sky, although one is a long way off and the other one much nearer. His object was to study the parallactic motion of the nearer star with reference to the more static distant one. But in investigating such apparent visual binaries, he discovered a considerable number of true visual binaries, i.e. pairs of stars that are closely related in space and not merely in direction, as seen from the earth.

In the year 1802 he found evidence of the orbital motion of a number of such double stars about each other, and two years later he wrote, 'Many doubles must be allowed to be real binary combinations of two stars intimately held together by the bond of mutual attraction.' Although, in this way, Herschel (and, following him, W. Struve) extended the range of Newton's law and shifted the main centre of attention of astronomers from the solar system to the stars, he never succeeded in his original object of measuring the distance of any star.

The Quest for Stellar Parallax

Some fifteen years after Herschel died, three other astronomers, the German F. W. Bessel, the Scot T. Henderson, and W. Struve, a German who had emigrated to Russia, within a few months of each other, in 1838 and 1839, all independently made the first reliable determinations of stellar distances. Even after that, progress was still very slow. Indeed, within the next fifty years hardly more than fifty stellar parallaxes were successfully determined. Now we know more than ten thousand.

As long ago as 1718 the proper motion of the stars had been discovered by Edmund Halley by comparison with observations made in antiquity of four conspicuous stars by Ptolemy, Timocharis and Hipparchus. The ancient observations were all in good agreement with one another. Halley said that these stars, Sirius, Aldebaran, Arcturus and Betelgeuse, 'being the most conspicuous in Heaven, are in all probability the nearest to the earth; and if they have any proper motion of their own it is most likely to be perceived in them, which in so long a time as 1 800 years may show itself by the alteration of their places, though it be utterly imperceptible in the space of a single century of years'. And, in fact, Halley deduced that these stars had moved in that period of time.

It may seem strange that the star whose distance was first determined was *not* one of the very brightest. Indeed, of the four just mentioned only Sirius is really a near star, the others being conspicuous primarily because of their high luminosity. But the luminosity or intrinsic brightness of a star must be distinguished from its apparent brightness. The former is independent of the distance of the star from the earth. The measure of luminosity is called absolute magnitude, that of apparent brightness apparent magnitude. All stellar magnitudes are measured on a geometric scale so that a difference of one magnitude corresponds to a ratio of brightness of approximately 2·512, or to be precise the fifth root of 100. Hence, to a difference of five magnitudes in measure corresponds an actual difference in brightness of 100 times. A first-magnitude star is 2·512 times as bright as a second-magnitude star, and so on, the higher the number of the magnitude the fainter the star. The conventional choice of zero-point for the scale is such that the apparent magnitude of Sirius, the brightest star in the sky, is about $-1·6$, whereas that of the sun is about $-26·7$.

The absolute magnitude is the magnitude which a star would have if it were brought to the standard distance from us of 10 parsecs. The parsec is the distance at which a star would have to be in order to have an annual parallax of one second of arc. On the scale of absolute magnitude, the sun's magnitude is 4·85, that of Sirius 1·3, and that of Rigel, the brightest star whose absolute magnitude has been reliably determined, about $-5·5$. Consequently, Rigel is intrinsically much brighter than the sun, in fact about 14 000 times as luminous; Sirius is only about 26 times as bright.

The first star whose distance was reliably determined, 61 Cygni, is comparatively faint, being only of the fifth apparent magnitude. It was chosen by Bessel for a parallax determination because the Italian astronomer Father Piazzi had discovered in 1806 that its proper motion was just over five seconds of arc a year. Because of this unusually large proper motion, 61 Cygni became known as the flying star. Only four larger proper motions have since been discovered, the largest being that of a faint tenth-magnitude star, Barnard's star, which has an annual proper motion of just over ten seconds of arc. Modern research has revealed that the range in observed stellar motions is less than the range in the intrinsic brightness or absolute magnitude of stars, and thus proper motion does, in fact, provide a much better general criterion of distance than apparent magnitude.

For a whole year Bessel measured the changes in the angular distance of 61 Cygni from faint 'neighbouring', but more distant, comparison stars, and at the end of 1838 he announced that, in the course of the year, 61 Cygni had described a very small ellipse in the sky, the image of the earth's path around the sun. The angle which the radius of the earth's orbit would subtend to an observer of 61 Cygni was found to be only three-tenths of a second of arc. Having measured this angle and knowing the distance of the earth from the sun, it was possible to calculate the distance of 61 Cygni, which is about 10 light-years (i.e. light travelling at 300 000 kilometres per second would take 10 years to cover the distance) or 3·3 parsecs.

Nowadays all direct determinations of stellar parallax are made photographically. A photograph of the star with reference to its more distant background stars is taken three times at six-monthly intervals. By comparing these three different pictures the proper motion of the star in space is disentangled from its apparent parallactic displacement due to the earth's annual revolution round the sun.

The nearest known star is Proxima Centauri, a companion of the triple system of stars known as Alpha Centauri. Its distance is just over four and a quarter light-years and its parallax a little less than one second of arc. If this three-star system is counted as one star, then there are some twenty-six stars within a distance of four parsecs, or about thirteen light-years, i.e. which show an annual parallax of more than a quarter of a second of arc. The approximate average distance apart of stars in our region of space is therefore of the order of five light-years. There are, however, probably several very faint stars still remaining to be discovered in our neighbourhood.

The Nearest Stars and the Brightest Stars

What are our stellar neighbours like? Of the ten nearest stars to us, counting the sun as one and also taking Alpha Centauri as being three stars, five belong to systems containing more than one star. Sirius, for example, consists of two stars. But only two of the ten nearest stars are of greater absolute luminosity than the sun: Sirius is 26 times as bright and Alpha Centauri 1·3 times. Six of these ten nearest stars have absolute magnitudes not greater than one hundredth that of the sun. Indeed, of the twenty-six nearest stars more than twenty are intrinsically very much fainter than the sun. Such stars are called dwarf stars.

The list of the twenty or so brightest stars, i.e. those of the greatest apparent magnitudes, is very different from that of the nearest stars. Only Sirius and two others appear in both lists. These twenty stars of greatest apparent magnitude include six of absolute magnitude more than 1 000 times that of the sun and one whose absolute magnitude is about 80 000 times that of the sun. Stars of such high intrinsic luminosity are called giants. The nearest, Betelgeuse and Spica, are about 300 light-years away, and it appears, therefore, that in our part of the stellar universe dwarf stars must be much more numerous than giants.

Stellar Luminosities, Masses, Temperatures and Diameters

Given the distance of a star and its apparent magnitude, its absolute magnitude can be easily calculated. The mass of a star, however, can only be determined if its pull on some other body is known. With the aid of the Newtonian theory of gravitation it is possible to measure the sun's pull on the earth and other planets. Similarly, the mass of a star can only be directly calculated if it is a component of a multiple system.

A surprising result emerges. Although the range of stellar absolute magnitudes is enormous, i.e. from less than one hundredth to more than one hundred thousand times that of the sun, the range in masses is unexpectedly narrow, only between one fifth and one hundred times that of the sun. Indeed, most stellar masses lie between 0·4 and 4 times that of the sun. For example, the mass of the brighter component of Sirius is 2·44 times the solar mass, whereas that of the fainter component is 0·96 times the solar mass, although the luminosity of the latter is only about one ten-thousandth that of the former.

The determination of the diameters of the stars is much more tricky since the stars, unlike planets, do not show up as discs in the telescope. With the most powerful telescopes even the nearest stars only give diffraction patterns producing spurious, and not true, discs. It was one of the great triumphs of modern observational astronomy when, in 1920, the angular diameter of a star was directly determined for the first time, after A. A. Michelson had suggested that his interferometer be applied to the 100-inch telescope on Mount Wilson. The technique was extremely difficult, but nevertheless the diameters of Betelgeuse and of several other giant stars were determined in this way. The diameter of Antares was found to be about 450 times that of the sun. If Antares were placed with its centre coincident with the centre of the sun, it would extend beyond the orbit of Mars and the earth would be right inside it! Betelgeuse, which also has a very large diameter, oscillates in size, the period of oscillation being about six years. During that time its radius waxes and wanes from about 210 to 300 times that of the sun.

The sizes of these stars are immense, but this was not unexpected. Their empirical determination provided a welcome check for what had been the only, and was to remain the principal, method for determining stellar diameters – a theoretical method based on the hypothesis that the stars could be regarded as perfect, or black-body, radiators. As in the case of the sun, this is not exactly true, but it proves to be a good working hypothesis. On this hypothesis, H. N. Russell of Princeton obtained a formula for the effective surface temperature of a star, involving a factor, known as the colour-index, which has to be determined empirically for each star. Given the effective surface temperature of a star and its absolute magnitude, its diameter can be readily calculated. The problem is simply to discover the size of a surface radiating at this temperature which would have the given luminosity.

The temperature of a star effectively controls what is known as its spectral class. Stars of different surface temperatures have different characteristic spectra. Effective surface temperatures above 50 000 K have been determined, and many bright stars have temperatures of 20 000 K and upwards, compared with the effective surface temperature of the sun which is approximately 6000 K.

Michelson's interferometric technique has been further developed since the nineteen-fifties with remarkable success by R. Hanbury Brown and R. A. Twiss. Recently, Hanbury Brown made an important advance by using this technique to measure the angle subtended by the relative orbit of the two components of the spectroscopic binary star Alpha Virginis. (A spectroscopic binary is a double star whose two components are so close together that their separate existence can only be deduced from study of their combined spectra.) Previous conventional spectroscopic observations yielded the period and velocity and hence the diameter of the relative orbit. This information combined with the angle measurement provides a new means of determining the *distance* of the system, which is found to be 275 light-years, to within five per cent. This result may be compared with previous estimates that ranged from 100 to 360 light-years.

The Three Principal Categories of Stars

Roughly speaking, most stars fall into three distinct categories: the main sequence stars, the red giants, and the white dwarfs. The main sequence stars form the leading category, the sun being an average intermediate member. They are, very roughly, of the same size – that is, the same diameter – as the sun. They are characterized by what is known as Eddington's mass-luminosity law, according to which the stars of greater mass are those of greater luminosity, those of smaller mass of fainter luminosity. If the spectral classes (temperatures or colours) are plotted against absolute magnitudes, these stars are found to lie within a fairly narrow band. This was first established in 1913 by H. N.

Russell of Princeton following a preliminary indication by E. Hertzsprung of Leiden. Main sequence stars are now recognized to be hydrogen-burning stars. The brightest have high surface temperatures and are blue or bluish-white in colour.

The other two categories are quite distinct. The red giants are stars of high luminosity, but comparatively low surface temperature. Consequently, they must be enormously large. The brightest and largest in this group, for example Antares and Betelgeuse, are called super-giants. The white dwarfs, on the other hand, are small stars of low luminosity, but high surface temperature. They are comparable in volume with planets rather than with the sun.

Since the range of stellar diameters so greatly exceeds the range of stellar masses, it follows that the range in stellar density must be very wide. The average density of the sun is little more than that of water, whereas that of Antares is less than one millionth that of water. On the other hand, the average density of a typical white dwarf is 100 000 times that of water.

Variable Stars

Not all stars are as steady in their behaviour as the sun, which exhibits only a very small degree of variability. Many vary in apparent brightness, the change being often periodic. This variation is sometimes due to external factors: for example, if the star is a component of a multiple system, its light may be eclipsed when it passes behind another star. There are, however, many intrinsically variable stars which may be roughly classified into pulsating and explosive variables. The former are, in the main, giant stars.

The periods of pulsating variables range from about one and a half hours to over a thousand days. Those with periods of less than a day form a very uniform group; they are called RR Lyrae variables after the type-star RR Lyrae. As these stars are believed to be all of very nearly the same absolute magnitude, they are very useful as distance indicators. Variables with periods exceeding one day are called 'classical cepheids', from the type-star Delta Cephei. The adjective 'classical' is added to distinguish them from the RR Lyrae variables, which are sometimes referred to as 'cluster-type cepheids' because they occur in the so-called globular clusters of stars.

The classical cepheids form a much less homogeneous group than the cluster type: they vary greatly in period, luminosity and spectrum. They are of much greater absolute magnitude; indeed, they include many of the brightest stars known. They have even been detected outside our own stellar system and have, therefore, proved to be very useful as long-distance indicators. In 1912 Miss Leavitt of Harvard College Observatory discovered an empirical relation between the average absolute magnitudes of these stars and their periods. Given the period of a known cepheid, its average absolute magnitude can be read off a graph or table, and hence it is possible to calculate its distance.

Explosive variables, or novae, are stars which increase tremendously and very rapidly in brightness; the increase in intrinsic brightness is between ten thousand and a million times and occurs within a period from one or two days up to two or three weeks. After that there is at first a fairly rapid and then a much slower decline. As far as is known, a star *usually* undergoes only one such nova outburst, but it must be remembered that the whole period of telescopic observation is very small indeed compared with the life-times of the stars.

The successive stages in a nova outburst are accompanied by drastic changes in the star's spectrum. As a rule, with the initial rise in brightness the spectral pattern not only changes in character, but moves bodily towards the violet, indicating the rapid expansion of the star. Indeed, the rate of such an expansion can be determined by this typical shift of spectral lines, which is known as the 'Doppler effect', because the measured shift is related to the motion in the line of sight due to the expansion of the side of the star facing the earth. After nova outbursts several novae have been observed to be surrounded by an expanding nebulous envelope, presumably formed by matter ejected from the star.

What is the cause of such violent stellar explosions? When the star is in a steady state, like the sun, there must be, at any point inside it, a balance between the gravitational pressure of the superimposed material and the pressures of gas and radiation. If there is a zone of instability for which such a balance may be easily upset, then a sudden extra liberation of energy in the star, due to some disturbance within it, may cause the star to make a violent transition to a new equilibrium state. Owing to the energy thus liberated, the gas in the star will become overheated, its pressure will rise as it tries to expand, and the overlying layers will be violently ejected into outer space.

Different views have been put forward concerning the kind of star which might be expected to become a nova and the possibility of the sun exploding in this way. Some forty years ago E. A. Milne suggested that every star must pass through the nova stage at some point in its evolution. On the other hand, D. B. McLaughlin has argued that only special stars become novae and that these stars may be novae more than once. It is difficult to choose between these two hypotheses. It is known that some stars have, in fact, been novae several times. Nevertheless, Milne's suggestion receives some support from the frequent occurrence of novae – in the Milky Way about ten nova explosions are observed in a year – and from a comparison with the number of stars believed to exist in our galaxy and the age assigned to it. If the sun suddenly became a nova all life on earth would perish rapidly.

There is a class of exceptionally brilliant explosive stars, called supernovae, which at maximum are from between ten to a hundred million times as bright as the sun. If a supernova appeared one hundred light-years away it would shine in the sky more brightly than the full moon. These stars are much rarer than ordinary novae: whereas in a stellar system such as the Milky Way ordinary novae occur at the rate of nearly one a month, it has been estimated from observing supernovae in external systems that one supernova occurs in our galaxy every four hundred years or so (see Plate XVI). Supernovae have been divided into two classes, one being on the average rather more than two magnitudes (absolute) brighter at maximum than the other.

We have historical records of only three supernovae which have occurred in the Milky Way. Two of these were Tycho's star of 1572 and Kepler's star of 1604, and both were of the second, or lesser, class. The only first class supernova in our galaxy of which we have definite record was observed by the Chinese in the year 1054, and there is now little doubt among astronomers that the Crab nebula (see Plate XVII), which is expanding, is its nebulous envelope. This has been confirmed from its position in the sky and its measured rate of expansion.

It is thought that, whereas in an ordinary nova outburst the outer layers of the star are blown off into space, a supernova explosion is much more prodigious, material being ejected not only from the outer layers but even from the deep interior. Spectroscopic investigations indicate that the initial velocities of ejection may be of the order of 5 000 kilometres per second. The Crab nebula is expanding at more than 1 000 kilometres per second and now occupies a space of about six light-years in diameter. In recent years this object has aroused considerable interest among astronomers because of the peculiar nature of its radiation. As shown in Plate XVII, when photographed with a colour filter which absorbs much of the light from the nebula but lets through certain wavelengths emitted by hydrogen and nitrogen atoms, it seems to consist of a multitude of filaments. If, however, it is photographed with filters which absorb all the stronger spectral lines of light ordinarily emitted by radiating atoms, we obtain a very different picture which is due to a continuous spectrum spread evenly over a wide band of wavelengths. This is very puzzling since light from extremely rarefied gaseous matter is normally confined to discrete emission lines in the spectrum. Astronomers believe that this unusual radiation is due not to atoms but to electrons moving at high speed in a magnetic field. This conclusion is based on the discovery in the laboratory that electrons accelerated to a very high velocity in circular paths by a magnetic field radiate light with a continuous emission spectrum (the synchrotron effect, see p. 76). The hypothesis that a basically similar mechanism operates in the Crab nebula has been strengthened by the discovery that, as in the case of the light from electrons accelerated in a synchrotron, light from the nebula is plane-polarized: that is, the light waves vibrate only in the plane perpendicular to the magnetic field. It is also believed that the synchrotron mechanism is responsible for the strong radio emission which has been observed to come from the direction of the nebula. Because electrons which emit on radio wavelengths have lower energies than those which emit on luminous wavelengths, they are likely to be more numerous, and this is confirmed by the fact that the radio emission is much the stronger. On the other hand, calculations have shown that the energy of the luminous electrons must be extremely high. It has also been estimated that explosions of the Crab nebula type could account for nearly ten per cent of the cosmic rays observed on earth.

Cosmic Rays and their Origin

The existence of cosmic rays was established early this century soon after the discovery of radioactivity, when it was observed that a small ionizing current could be detected even in the absence of any radioactive source. In 1912 it was established that the radiation producing this current could not be of terrestrial origin. A young Austrian physicist, Victor Hess, measured the intensity of this radiation at various altitudes in a balloon and found an increase of intensity with height above sea-level. He concluded that the radiation must enter the atmosphere from above. This view was confirmed by later investigators, notably by Millikan and Cameron in 1928 who found that at successively greater depths under water the radiation decreased. Millikan gave the radiation the name *cosmic rays*.

At first cosmic rays were believed to be photons (hence the use of the word 'rays'), but when it was found that their intensity varies with latitude, being greatest in the neighbourhood of the earth's magnetic poles, it was realized that this was impossible because photons have no electric charge and are in no way affected by a magnetic field. The geomagnetic effect has been studied in great detail and confirms the assumption that the primary 'radiation' comes from outside the earth. This 'radiation' arrives fairly uniformly from every direction in space and consists mainly of particles of very high energy, most being of the order of 10^9eV.

The cosmic rays observed at the earth's surface are secondary particles, products of collisions between the primary particles and the molecules of the atmosphere. The primary particles are mostly atomic nuclei, particularly protons. The relative abundances of the various atomic species in cosmic rays are more or less in accord with their relative abundances in the universe generally.

The total energy density of cosmic rays in the vicinity of the earth is about 10^{-19} joules per cubic centimetre. This is comparable with the density of starlight. Above 10^9eV the relative number N of particles with energy greater than E is given by the empirical formula

$$N = \frac{\text{constant}}{E^n},$$

where $n = 1 \cdot 6$ for values of E up to 10^{15}eV, but between 10^{15}eV and 10^{18}eV the value of n changes to $2 \cdot 2$. It may be that this change corresponds to a transition from galactic to extragalactic sources. Above 10^{14}eV the sparsity of particles is such that they can only be studied through the cascade showers of secondary particles that these very energetic primaries produce. A few particles have been observed with energies in excess of 10^{20} eV.

Theories of the origin of cosmic rays assume that the main source lies outside the solar system. Otherwise, we would not observe the intensity to be the same in all directions. Furthermore, if most cosmic rays emanated from the sun, we would expect some correlation between cosmic ray intensity and solar activity; but, although large flares on the sun often cause an increase in the cosmic rays on earth, this effect is far too small to be of much significance. Moreover, even during violent flares the most energetic particles emitted by the sun have energies not greater than 2×10^9eV. Consequently, the principal sources of cosmic rays have been sought in our stellar system and even beyond.

One of the most important suggestions for explaining the acceleration of cosmic ray particles inside the galaxy was made by Fermi in 1951. It was based on the discovery by Hall and Hiltner in 1949 that light from some space-reddened stars, i.e. stars that appear to be reddened due to intervening material, is also partially plane-polarized. This was interpreted as being due to the effect of a general galactic magnetic field. Previously, there had been no evidence that magnetism featured as an effective galactic force. Fermi suggested that, provided the particles had fairly high velocities originally (ejection from stars might account for that), the interstellar magnetic field could accelerate them up to the very high velocities observed and also bend their paths round and round, thereby trapping them within the Milky Way and causing them to impinge on the earth equally from all directions. This ingenious theory, which yielded the correct form of law for the dependence of N on E, has now been abandoned as an explanation of high accelerations, for detailed analysis has shown that interstellar space is incapable of increasing the energy of a particle by more than a factor ten at most. Instead, it is now generally believed that the main role of interstellar space as far as cosmic rays are concerned is to act as a diffusive mechanism for smoothing out the original directions of motion so as to yield an isotropic flux, i.e. one that is the same in all directions, and also for producing the observed law of the dependence of N on E.

As for the ultimate origin of cosmic-ray particles, it is now generally believed that those with energies exceeding 10^{16}eV must come from outside our galaxy, and that only those of lower energies can be produced by localized sources within it. As previously mentioned, supernovae explosions of the Crab nebula type may perhaps account for nearly ten per cent of the cosmic rays observed on earth, the absence of electrons from these rays being due to their loss of energy by the process of synchrotron radiation in magnetic fields. Ordinary novae may also make a significant contribution, particularly at the lower energies. We shall return to the intriguing question of the origin of the most energetic cosmic rays later (see pp. 275–6).

An unexpected effect due to cosmic rays has been observed on space flights. The peculiar flashes of light first seen in 1970 by the Apollo astronauts, even when their eyes were closed, were probably caused by the direct reaction of cosmic ray particles with the constituent atoms of the fluid inside the eye. This produced a shower of electrons triggering optical receptors on the retina.

Pulsars

Towards the end of 1967 one of the most exciting astronomical discoveries of the century was made by A. Hewish and his colleagues at the Mullard Radio Astronomy Observatory at Cambridge. In the summer of that year a new radio transmitter, designed by Hewish to operate at a frequency of 81·5 MHz (3·68 metres) for the study of radio sources of small angular size, came into service. These sources give rise to a phenomenon known as 'interplanetary scintillation' when the radiation from them passes through plasma clouds ejected by the sun. The time scale of this scintillation is of the order of a fraction of a second. In August 1967 Jocelyn Bell, a research student working with Hewish, found in addition to the expected scintillating sources, a new fluctuating signal observed in a direction well away from the sun where no scintillation had been expected. Later that year these signals were found to be pulses of variable strength but all less than one tenth of a second long, occurring in a regular sequence at intervals of just over a second.

At first it seemed that these signals might be from intelligent beings on some planet near another star, but the variations in strength proved to be random and it was soon clear that the source was natural and not an artifact. Three other pulsating sources, one with a period of a quarter of a second were soon found, and the term 'pulsar' was introduced. The most striking feature was the extreme regularity of the pulsations, the period of the first to be discovered being precisely 1·33730113 seconds. Later it was found that although some pulsars keep time to an accuracy of 1 part in 10^9 over a period of weeks, there is a systematic slowing down.

Although the precise nature of pulsars is still a mystery, the regularity of their radio pulses indicates that they are oscillating or rotating bodies. In either case they must be very small, because no body can emit a sequence of coherent pulses of radiation in a time shorter than would be required for light to travel across it. It was therefore suggested that pulsars might be white dwarf stars in a state of rapid pulsation or rotation, but the discovery of some with periods of the order of a tenth of a second made this view untenable because their densities would be too high. The diameter of a pulsar with a pulse interval of a tenth of a second cannot exceed 30 000 kilometres, which is comparable with the size of the earth rather than the sun. Moreover, in any pulsation or rotation of a massive body the shortest possible period is controlled by gravity and is of the order of $1/\sqrt{(G\rho)}$, where G is the universal constant of gravitation and ρ the density. Consequently, a period of one tenth of a second can only be achieved by a density of the order of 10^9 grammes per cubic centimetre, which is greater than that of a white dwarf (10^6 grammes per cubic centimetre).

In a white dwarf star the constituent atoms have lost their attendant electrons and are reduced to tightly packed nuclei with electrons floating freely through the whole mass. The gravitational force tending to compress the body further is balanced by the pressure arising from this state of electron 'degeneracy', but there is a limit to this balance. At densities of about 10^9 grammes per cubic centimetre further compression does not generate sufficient pressure to balance the increased gravitational attraction, since the electron population is decreased by inverse β-decay. A star of this density therefore continues to contract until the nuclei of its constituent atoms are disrupted and the material consists only of neutrons packed tightly together, with a few remaining protons and electrons moving freely around. At densities of from about 10^{13} to 10^{16} grammes per cubic centimetre, however, pressure and gravity can be in balance again. If such dense matter could exist on earth a teacupful would weigh hundreds of millions of tons, and if the earth itself were compressed to this density its diameter would be only about a hundred metres! A star of this density would have a radius of the order of 10 kilometres.

The properties of stars of this type, called *neutron stars*, were first investigated theoretically by Landau (in U.S.S.R.) and Oppenheimer and Volkoff (in U.S.A.) in the nineteen-thirties, but until the discovery of pulsars they remained merely a hypothetical possibility. In attempting to construct a pulsar model based upon the idea of a neutron star the first difficulty, oddly enough, is to obtain a slow enough period, for if a neutron star retains the angular momentum typical of an ordinary star it must spin at roughly 100 to 1 000 cycles a second, rather than at 1 to 10 cycles. This difficulty was removed when it was discovered that, although the most rapid pulsars are slowing down at such a rate that the period of rotation would double in a few thousand years, in the case of slower pulsars a doubling of the period would require tens of millions of years. On this basis, it would appear that in the early stages of their careers pulsars may in fact rotate at about a thousand cycles a second, but this phase is so short-lived that the chance of finding so rapid a pulsar is very small.

Although at present there is no general agreement about the precise mechanism involved, most astronomers believe that pulsars are probably rotating neutron stars. For, if rotation regulates the timing and also provides the energy source of the radiation emitted, the observed increase of period indicates the rate of loss of energy, and even for those pulsars that exhibit the slowest variation in period the deduced loss of energy is sufficient to sustain the observed radio emission. If the pulses are to be explained by a lighthouse type of model it is necessary that the radiation be confined to a beam of about 15 degrees' width.

Various suggestions for the pulsar mechanism have been made, all based on the plausible assumption of a strong magnetic field. For, since most stars are thought to have magnetic fields and these fields increase on compression (if the decay time of the field is long compared with the compression time of the star), it is likely that an intense field, perhaps as high as 10^{12} gauss, may exist at the surface of a neutron star. In such a strong field any ionized gas that escapes from the surface will tend to move along the magnetic lines of force and so will be whirled around as the star rotates. In 1968, T. Gold suggested that this is what in fact happens in a pulsar. He argued that plasma escaping from the star is swept into co-rotation by the magnetic field and that at distances of a thousand to two thousand kilometres its velocity will approach that of light. The plasma will then radiate by the synchrotron mechanism if there are sufficient variations in its charge density. Because the radiating plasma is moving at nearly the speed of light, the emitted signals will be strongly beamed in the forward tangential direction so that to a distant observer the neutron star will behave like a lighthouse, the duration of each flash depending on the angular width of the rotating sector of plasma. At greater distances the plasma will no longer adhere to the constraining field but will tend to stream outwards.

Among other models suggested, mention may be made of F. Pacini's idea that a rotating neutron star with an oblique magnetic axis could emit sufficient energy as a classical dipole radiator oscillating at the rotation frequency to account for the pulsar phenomenon. V. L. Ginzburg, on the other hand, has suggested that a stream of plasma flows outwards along the magnetic lines of force from each pole of the neutron star. Plasma instabilities develop and radio waves are emitted at right angles to the stream. By arranging a suitable orientation of the magnetic axis, more than one flash per revolution could occur and this might account for some of the inter-pulse phenomena that have been observed.

Information about the distances of individual pulsars is still very incomplete, but the distribution of the fifty or so now known shows a small but significant concentration towards the main plane of the Milky Way indicating that those observed are at distances not greater than the thickness of our galaxy, which is about 1,000 light years. Attempts to identify pulsars with known objects were disappointing until, late in 1968, a southern hemisphere pulsar was found to be associated with a radio source known as Vela X, believed to be the remnant of a supernova explosion. Its period was 89 milliseconds. Shortly afterwards a pulsar was found at the heart of the Crab nebula with a period of 33 milliseconds making it the fastest and possibly the youngest pulsar known. Then in January 1969 optical pulsation of the same rate was discovered in this nebula, the source being the star that had been regarded for many years as the probable origin of the supernova explosion. What had not been previously realized was that all its light is emitted in short-duration flashes (see Plate XVIII). More recently it has been found that the same star is emitting pulsed X-rays and γ-radiation.

In September 1969 the frequency of the pulses of the Crab pulsar suddenly and unexpectedly *increased*. This event was attributed to either the effect of a starquake causing a change in the structure of the star or to disturbances produced by a hypothetical orbiting planet. Whatever the explanation, it was found in the course of 1970 that the variation of the pulses seemed to be associated with changes in the strength of a spinning magnetic field. This was the first direct indication that the spinning magnetic field proposed by theoreticians to account for pulsar radiation actually exists.

In the spring of 1971 it was announced that N.A.S.A.'s X-ray explorer satellite had revealed the existence of an X-ray pulsar associated with the X-ray source Cygnus X-1, which has been known for some years. This discovery was surprising because the other properties of this source are very different from the only other X-ray pulsar known – the one in the Crab nebula. In Cygnus X-1 there is no evidence of radio pulses, nor any visual evidence of a surrounding nebula typical of a supernova remnant, and as yet no evidence of optical pulsation. The rate of X-ray pulsation, however, is 15 pulses a second, suggesting a relatively young age of about 10,000 years. It has been suggested that if Cygnus X-1 is not a rotating neutron star it may be a still denser type of body, the pulsed X-rays coming from a cloud of matter orbiting around it and spiralling in towards it.

Black Holes

Are there any objects denser than neutron stars? Bodies of density greater than 10^{16} grammes per cubic centimetre cannot exist in the form of neutron stars and their collapse to infinite density appears to be inevitable. So long as white dwarfs were the densest objects known, the question of complete gravitational collapse seemed to be unreal, but now that the evidence for the existence of neutron stars is so strong, we view the problem in a very different light.

A 'black hole' is the name that has been given to what is left behind after a body has suffered complete gravitational collapse. To study such an object we must invoke Einstein's general theory of relativity, or else envisage some modification of that theory at very small distances. For example, it has been suggested recently that at such distances gravitation may cease to be an attraction and become repulsive. If, however, we stick to general relativity, then some startling consequences follow when the radius of a body shrinks to less than a certain value depending on its mass: $2Gm/c^2$, where m is the mass, G the universal constant of gravitation, and c the velocity of light. For a star such as the sun, this 'Schwarzschild radius', as it is called, is about 3 kilometres. (If its radius were to shrink to this size, the density of the sun would be nearly 2×10^{16} grammes per cubic centimetre.) From a body that has shrunk to its Schwarzschild radius, no radiation can emerge nor can any matter be ejected. Consequently, to an observer outside the body it becomes completely invisible and no information from it is obtainable. On the other hand, to an observer located on it complete collapse of the body to zero radius and infinite density *in a finite time* is inevitable.

That a sufficiently massive body of cold matter will necessarily collapse to a black hole was first established on the basis of Einstein's theory of gravitation by Oppenheimer and Snyder, in 1939, on the assumption of spherical symmetry. In other words, they neglected rotation. What happens in the case of rotation is much more difficult to analyse.

Of all objects that may exist in the universe none would seem to offer a poorer prospect of detection than a solitary black hole of stellar mass. Consequently, the only hope of detecting the existence of such a body would be by its effect on some near-by body, in particular if it were a member of a binary system. In the case of close binary stellar systems a flow of matter from one component to the other is well known from theoretical analysis of spectroscopic data. Gas being funnelled down a black hole from a companion star would undergo heating by compression to temperatures of 10^{10} to 10^{12}K, but only a small part of its radiation would escape. Ya. B. Zel'dovich and J. D. Novikov have calculated that most of this radiation would emerge either in the visible part of the spectrum or in the X-ray or γ-ray region, depending on the mass of the black hole.

Early in 1971, A. G. W. Cameron drew attention to the eclipsing binary Epsilon Aurigae, the primary component of which is a supergiant star of mass 35 times that of the sun. The other component cannot be seen directly but radiates in the infra-red. When it passes in front of the primary it cuts off about half the total light we receive. A total eclipse lasts about 700 days, but during the central 330 days the light we receive remains constant at its minimum value. From this it has been calculated that the diameter of the secondary is over 2 000 million miles, i.e. greater than the diameter of the orbit of Uranus about the sun. This would make it the largest star known, and Cameron has therefore suggested instead that it is a black hole surrounded by a cloud of solid particles (from which the observed infra-red radiation comes) orbiting at some 1 500 million miles from the centre of mass of the system. However, this interpretation has been contested and it remains to be seen whether the existence of any black hole has been – or will be – incontrovertibly established.

Galactic Nebulae

In the course of his observations of the Milky Way Sir William Herschel discovered a number of objects which looked like planets but showed no typical planetary motion. He eventually assigned to them the name ' planetary nebulae'. These nebulae are inside our own system and must be distinguished from those external nebulae which are more appropriately called galaxies to emphasize their similarity with the Milky Way, or the galaxy.

When a planetary nebula is observed with a sufficiently power ful telescope a faint star is usually detected at the centre. Some planetary nebulae are ring-like and their distances vary from 3 000 to 30 000 light-years. The nebulous shell emits light owing to excitation by the central star, which must, therefore, be very hot, its effective surface temperature being from 50 000 to 100 000 K. Although the diameters of planetary nebulae are about ten thousand times the distance of the sun from the earth, their masses are less than a fifth that of the sun. They are really great glowing near-vacua thousands of times rarer than the best vacua obtainable on earth. We see them only because they are so large.

The strongest lines in the spectra of these nebulae are certain green lines which for long resisted identification. It was even suggested that perhaps they were due to an element not known on earth, to which the name 'nebulium' was assigned. However, in 1927, I. Bowen found that these mysterious lines were due to doubly ionized oxygen, and that the conditions which are necessary for producing them, namely very low density of gas and exposure to exceedingly weak radiation, are impossible to obtain in the laboratory. Only the great extension of the materials in which these conditions occur enables a strong line to be built up in the spectrum.

Although the central star must be very hot, it is usually rather faint. Paradoxically, although the light from the nebulous envelope can ultimately be traced back to the star, it may be as much as fifty times that emitted by the star itself. This strange effect is believed to be due to fluorescence. The star at the centre is so hot that most of its energy is radiated in an invisible part of the spectrum, in the far ultra-violet. The nebula absorbs this ultra-violet radiation, but re-emits it as visible light. The high surface temperature of the central star, combined with its low absolute magnitude, shows that it must have a small surface area. This conclusion suggests comparison with the white dwarfs. Milne thought that planetary nebulae might have resulted from former novae, but as only about one hundred and fifty planetary nebulae have so far been identified they are too few to suppose that every nova gives rise to such a nebula. The whole problem is still very obscure. The planetary nebulae tend to have definite shapes which are believed to be well maintained by rotational motion about their centres.

Far more numerous in the sky are the amorphous diffuse nebulae in our stellar system. These are of two kinds, the bright and the dark. Bright diffuse nebulae appear to consist of glowing clouds and wisps of matter in chaotic motion (see Plates XIX and XX). Hubble's investigations at Mount Wilson some years ago showed that in almost all cases the light from such a visible diffuse nebula is due to some associated star or stars. These stars are like beacons shining upon a mixture of fluorescent gas (atoms) and reflecting dust (particles of diameters about one hundred-thousandth of a centimetre, comparable with the wavelength of visible light). When the associated star is very hot, having an effective surface temperature of about 50 000K, the gas fluoresces and an emission spectrum with bright lines is obtained. When the associated star is cooler, its effective surface temperature being less than 18 000K, the spectrum shows dark absorption lines. In this case the starlight is scattered by the particles and the spectrum of the nebula is simply a reflection of the stellar spectrum.

Generally, the association between a bright cloud of dust and gas and a star or stars is merely cosmographical and not physical. If there is no star in the neighbourhood, the nebula appears as a dark patch in the sky, absorbing and scattering light from stars a long way off. Herschel was interested in such dark patches in the region of the Milky Way and wondered if he was penetrating into the depths of space beyond. Another hundred years or so went by before, towards the end of the nineteenth century, E. E. Barnard of Yerkes showed conclusively that such dark patches were really obscuring clouds of dust.

Luminous and dark nebulae are often in close association (see Plate XXI). Dark nebulae are, as a rule, far more extensive than bright ones. Nebulae contain atoms, molecules, dust particles and larger particles, but dust is the main obscuring agent, and its relative amount may vary widely. Indeed, if a given mass of material is to be converted into the most effective obscuring agent possible, it must be split up into particles of this size.

Interstellar Matter

Most of these obscuring clouds are found in the region of the central band of the Milky Way. Besides these concentrations of dark matter, there is, however, a great deal of more diffuse and rarified obscuring matter in and near the main plane of the Milky Way.

The discovery of interstellar gas goes back to an observation made in 1905 by Hartmann who detected in the spectrum of a certain star in Orion a curious spectral line which differed from the rest. This line was sharp and distinct, whereas the others were all fuzzy, presumably due to the effect of two close stars revolving around each other. In 1909 it was suggested that the sharp line was due to the effect of calcium vapour lying between the star and us. Not until 1924 was it generally realized that this calcium vapour was not an envelope surrounding the star but was truly interstellar. Recent investigators have all come to the conclusion that, in fact, the most abundant interstellar element is hydrogen, but since the interstellar absorption lines of hydrogen, unlike those of sodium and calcium, occur principally in the far ultra-violet, they are not visible. It has been calculated that the mean density of this interstellar gas is about one gramme per 10^{24} cubic centimetres, which is roughly about one hydrogen atom for each cubic centimetre. This is less than one part in a million of the density of the material in, say, the Orion nebula (shown in Plate XX).

The light from distant stars seen through dust clouds is reddened and dimmed, and consequently stellar distances derived from measures of apparent brightness have had to be corrected. In the nineteen-thirties investigation of this 'extinction' of starlight revealed that the intensity of light is diminished by a factor of about ten for every thousand light-years of travel in interstellar space, and that the main source of this dimming was particles of radii comparable with optical wavelengths. The total dust density was found to be about one per cent of the smeared-out density of stars and interstellar matter in the neighbourhood of the sun.

Until 1964 it was generally thought that interstellar dust was composed of ice, solid methane and solid ammonia, but when investigation of the interstellar extinction curve in the far ultra-violet part of the spectrum became possible with equipment carried in rockets and artificial satellites a serious discrepancy between theory and observation was revealed. Further evidence against the hypothesis of ice particles came when dust was discovered in regions of ionized hydrogen near hot stars where the gas temperature is of the order of several thousand degrees. Moreover, infra-red radiation corresponding to temperatures of about 500°C was detected from dust particles near hot stars, and a much more refractory material than ice was therefore indicated. The first such material to be considered was solid carbon, in the form of graphite. F. Hoyle and N. C. Wickramasinghe, in Cambridge, showed theoretically that graphite particles of the required size could condense in the atmospheres of cool carbon-rich stars and be ejected into interstellar space by the pressure of stellar radiation. However, in order to account for the extinction data at ultra-violet wavelengths, it has since proved necessary to invoke some additional mechanism and particles of iron and magnesium silicate ejected by supernovae and certain oxygen-rich stars have been suggested.

The existence of simple diatomic molecules like CH and CN in interstellar space has long been known from the appearance of their characteristic absorption bands in stellar spectra. In recent years many others have been detected from spectral features in the microwave regions, the first so found being the hydroxyl radical OH at wavelengths close to 18 centimetres. The presence of interstellar molecular H_2 has also been observed. It is now thought that dust grains, formed in certain stars and supernovae and then ejected into space, could provide solid surfaces on which interstellar atoms can be absorbed and recombine to form molecules.

The main outcome of recent work in this field has been to show that, besides its role in the extinction of starlight, interstellar dust probably plays an important part in the processes of star formation and decay. It has been suggested that, when silicate grains in the deep interiors of dark clouds are cooled to about 3 K, they act as condensation centres for solid hydrogen. Once freezing on the grains has begun, it probably causes fragmentation of the gas clouds leading eventually to the formation of new stars. Eventually these stars either cool or explode, producing more dust and thereby starting the cycle all over again. It has been calculated that the time required for such a cycle is roughly a thousand million years, and that in the life of our galaxy about ten of these cycles have been completed.

Globular Clusters and the Size of the Milky Way

Another useful test for the occurrence of interstellar dust and gas is provided by a different class of stellar objects called globular clusters. These are great spherical conglomerations of stars (see Plate XXII). About one hundred have been identified surrounding the Milky Way. Their masses are thought to be of the order of one hundred thousand times that of the sun. It was observed that the light from the more distant globular clusters was reddened and that this reddening increased with distance. The observations were consistent with the hypothesis that the reddening was due to the passage of light through interstellar dust, just as the sun looks redder when seen through mist.

The globular clusters are important for the study of the Milky Way because cluster-type cepheids, which are useful as distance indicators, have been identified in them. More than fifty years ago H. Shapley, with their aid, was able to calculate the distance of the clusters, and he came to the conclusion that they form a system surrounding the Milky Way and concentric with it. Their peculiar distribution in the sky – they are nearly all confined to one-half of the celestial sphere – led him to conclude that the centre both of the system of globular clusters and of the main body of the Milky Way could not be somewhere near the sun, as had previously been thought, but must lie about thirty thousand light-years away in the direction of the rich star-cloud in Sagittarius where the Milky Way is thickest.

As the diameter of the Milky Way in the main plane is now estimated to be about eighty thousand light years, the sun is very far from being central. The greatest thickness of the Milky Way in the direction perpendicular to the main plane is about one fifth of the diameter of the main plane, i.e. about sixteen thousand light-years, but there is no truly precise boundary, only a falling-off in the number of stars. The diameter of the spherical system of globular clusters is about one hundred and thirty thousand light-years.

The Rotation of the Milky Way

This picture of our galaxy, which was very revolutionary when first put forward, has received independent support from the study of its rotation. This phenomenon was conclusively demonstrated by the Dutch astronomer Oort in 1927. If the whole system is rotating under its own gravitational field about its centre, and if the main mass of the Milky Way is concentrated towards this centre, then the general motion of the stars in the outer regions should be similar to the motion of the planets about the sun. The outer planets tend to lag behind the inner ones, and similarly in the stellar system the stars which are farther from the centre should lag behind those that are nearer. Hence, stars farther from the centre than the sun should lag behind the sun in this general motion around the centre and those that are nearer should race ahead. By studying the actual motions of the stars and analysing them statistically with this idea in mind, Oort was able not only to show that the Milky Way rotates but also to obtain a general confirmation of Shapley's result that the centre of the system was about 30 000 light-years from the sun.

Oort's method went further than Shapley's because, being based on the theory of gravitation, it led to an estimate for the total mass of the whole system. The orbital speed of the sun around the centre of the Milky Way was discovered to be about 220 kilometres a second. Hence, the sun makes a complete rotation around the centre of the Milky Way in about 225 million years. From this result the total mass of the system is estimated to be between one and two hundred thousand million times that of the sun. At one time it was thought that this mass was divided more or less equally between stars and diffuse matter, but we now believe that only about one or two per cent is in the form of interstellar gas. The total number of stars in the Milky Way is thought to be of the order of 100 000 million.

Galactic Clusters and Expanding Stellar Associations

Besides the globular clusters there are many other groups of stars in our galaxy in which the star density is higher than in the general star field surrounding them. Each such group moves as a whole around the centre of the galaxy, and at the same time its members move about its centre of gravity. Whereas globular clusters contain tens of thousands of stars and are found both near the plane of the Milky Way and at great distances from it, the open, or galactic, clusters contain tens or hundreds of stars but are far less regular in structure and are found only near the plane.

Whereas globular clusters are highly stable systems which cannot easily be disrupted, galactic clusters, particularly those in which the star density is relatively high, can be disrupted in periods of time of the order of 1 000 million years. Close encounters between individual members of such a cluster will occasionally result in a star acquiring so much energy that it can overcome the gravitational field and leave the cluster. In this way the system is slowly impoverished. Moreover, this disintegration process has been shown to be effectively irreversible. If clusters cannot be formed from stars already existing, it follows that the stars in a galactic cluster must have been formed simultaneously with the cluster.

About 350 open galactic clusters are now known. In addition there also exist in or near the main plane of the Milky Way loose clusters of hot stars known as *expanding stellar associations*. About 80 are known but since, like open clusters, they lie in regions occupied by interstellar matter, many others must be obscured from us and there are probably several thousand in all. Although their existence had previously been known, interest in these associations was greatly stimulated after the last war by V. A. Ambartsumian who pointed out that, under the tidal forces produced by the general gravitational field of the galaxy, they tend to disintegrate much more rapidly than more compact clusters and are completely dissipated in times of the order of a few million years. Consequently, the fact that they exist at all shows that these associations must consist of very young objects. Moreover, detailed study of their motions shows that the component stars of an association tend to recede from a common centre; for example, if we extrapolate backwards the motions of the stars in the Zeta Persei association, we find that it appears to have an age of about one and a half million years. The formation of stars is therefore still going on in our galaxy, and arguments have been advanced that all stars recently formed or now in process of formation are members of stellar associations. It seems that stars are being formed in groups which then tend to expand and disintegrate.

The Spiral Structure of the Milky Way

The solar system is situated in a very dusty region of the Milky Way, and we cannot, therefore, see very far in any direction in the main plane. Since we are also on the outskirts of the system, it is as though we were attempting to study the lay-out of London on a very foggy day from the roof of a building in the suburbs. The central core of the galaxy, although believed to be in the direction of the rich star-cloud in Sagittarius, is largely concealed from our vision by dust clouds. These clouds, however, are much less effective in obstructing the passage of radio waves whose wavelengths are from about one thousand to ten thousand million times those of visible light. The development of radio astronomy in the last decade has, therefore, provided a powerful new technique for investigating the general structure of the Milky Way.

About 1944 Van de Hulst, a young Dutch astronomer, predicted theoretically that a well defined spectral line associated with the neutral hydrogen atom existed in the radio range and, moreover, that it should be readily detectable under the physical conditions prevailing in interstellar space. Each of the two electric charges (proton and electron) which form this atom spins like a top, and so gives rise to a small magnetic field. These two fields can point in the same or in opposite directions. According to Van de Hulst, if they point in the same direction then, on the average, after some millions of years the atom will spontaneously switch over to the state in which they point in opposite directions, at the same time emitting radiation of approximately 21 centimetres in wavelength. This line has since been found observationally and has yielded valuable information on the irregular distribution of the interstellar gas, which mainly consists of hydrogen, and hence of the structure of the Milky Way. This investigation indicates that, as had long been suspected on other grounds, our galaxy is of spiral shape with arms which trail behind as the whole system rotates about its central core. However, for a fuller exposition of present ideas concerning the physical significance of spiral arms we must turn away from the study of our own stellar system and survey regions of the universe beyond the Milky Way.

CHAPTER 11

The Extent of the Universe

Beyond the Milky Way

Already in the eighteenth century, Wright and Kant, in their speculations on the structure of the physical universe, looked beyond the Milky Way. Wright suggested that just as there are other suns besides our own, so there are other galaxies besides the one of which the solar system is a part. Kant developed this conception further. He imagined other systems of stars, each being so far away from us that even with the telescope we cannot distinguish their components. He argued by analogy with the Milky Way that such a stellar world would appear as a faint spot, circular in shape if its plane were perpendicular to the line of sight, and elliptical if seen obliquely. Observational astronomers were already aware of such objects in the heavens and had called them nebulae without knowing whether they were in fact beyond the Milky Way. Here, to quote Kant's own words, 'a vast field lies open to discoveries', to which, he added, 'observation alone will give the key'.

The pioneer of extragalactic exploration was William Herschel with his great reflecting telescope made by his own hands. Then came Lord Rosse with his famous telescope made in Ireland about the time of the potato famine in the 1840s; and finally, in our own day, the astronomers in California with the 100-inch and 200-inch telescopes and all the modern photographic and other ancillary instruments and techniques.

The problem of the nature and status of the nebulae was, perhaps, the most difficult which Herschel tackled, particularly when we recall that he never succeeded in determining the distance of any single body outside the solar system. The complexity of the problem for the pioneer investigator is the more evident when we realize that the term 'nebulae' was originally applied equally to the glowing masses of diffuse material inside our own system and to the globular clusters which surround it, as well as to the other systems which lie outside our own in the depths of outer space. It is not surprising that until some thirty years ago man's ideas concerning the nebulae were characterized by quite as much uncertainty and controversy as still persist today on the even larger question of the universe as a whole.

In 1785 Herschel came to the conclusion that all nebulae were unresolved aggregations of stars outside our own system, thus imagining that he had penetrated the boundaries of the Milky Way. Over thirty years later, in 1817, he admitted that the utmost stretch of the space-penetrating power of his telescope could not fathom the profundity of the Milky Way. Although he still believed that some distant nebulae were independent stellar systems, the evidence appeared to be confusing and inconclusive.

Nearly half a century was to elapse before one of the great pioneers of astronomical spectroscopy, Sir William Huggins, found that the light from the Orion nebula and some others was similar to that of a glowing mass of gas. Consequently, observational evidence then favoured the view that all unresolved nebulae were merely glowing clouds of gas inside the Milky Way.

Meanwhile in 1845 Lord Rosse had set up his famous reflecting telescope with its six-foot mirror at Birr Castle in the centre of Ireland. Although such an instrument proved on the whole to be ineffective for the investigation of objects beyond the Milky Way (because, as we now know, these distant regions can only be studied satisfactorily with the aid of photographic methods involving long exposures rather than by visual observers however acute their seeing ability), Lord Rosse was rewarded by one of the great discoveries of observational astronomy. Within a few weeks of its completion, his six-foot mirror revealed for the first time that the spiral form, so lavishly employed by nature in the organic world, also occurs in the heavens. The Whirlpool nebula (see Plate XXIII) was the first spiral nebula to be recognized as such. Lord Rosse ultimately discovered fourteen similar objects and since then many more have been revealed with improved telescopes and long photographic exposures. By 1918, when the 100-inch reflector was erected on Mount Wilson, it was already estimated that the number of spiral nebulae visible in the heavens must be at least half a million. Nevertheless, despite their enormous number and peculiar structure, all these objects were still thought, particularly by the more cautious observers, to be constituents of the Milky Way. Six years later this conservative hypothesis was to be discarded for ever.

The problem was finally solved by identifying bright stars in the Andromeda nebula. As long ago as 1885 a star had been identified in that particular spiral. In August of that year a new star suddenly appeared in the central region and soon attained a luminosity of about one tenth of that of the whole nebula. Its position in the nebula and its spectrum, which was quite different from that of a typical nova, showed that it was not a foreground star. In 1917 two other novae were discovered from a study of photographs taken of the same nebula, but they were thousands of times fainter than the object seen thirty-two years before. It followed that if the 1917 objects were typical novae, then the Andromeda nebula must be of the order of a million light-years away and of a size comparable with the whole Milky Way. But if, on the other hand, the 1885 object was a typical nova, then this nebula was comparatively small and near.

Controversy raged until 1924 when, with the 100-inch telescope, Hubble succeeded in identifying cepheid variables in the Andromeda nebula. As mentioned in the previous chapter, cepheid variables which are among the stars of highest absolute magnitude had been studied in 1912 by Miss Leavitt, who found an empirical law correlating the average luminosities of these stars and their periods, from which their relative distances can be calculated. But it was still necessary to determine in some other way the distance of one cepheid variable in order to fix the scale absolutely. This was not easy since the nearest cepheids to us in the Milky Way are rather remote; Delta Cephei, the type star, is several hundred light-years away.

On the basis of the absolute scale which was eventually constructed, Hubble assigned a distance of about 900 000 light-years, which was later scaled down to about 750 000 light-years, to the

Andromeda nebula. (As is explained on p. 253, the Andromeda nebula is now believed to be more than twice as far away as Hubble estimated.) Since the Milky Way has a main diameter of about 100 000 light-years, Hubble's result showed conclusively that the Andromeda nebula was an independent stellar system of roughly comparable size. Incidentally, this also confirmed the existence of supernovae as objects distinct from ordinary novae. The star observed in the Andromeda nebula in 1885 must have been far more cataclysmic than any ordinary nova, attaining a much greater absolute brightness. Indeed, a supernova can itself become as bright as the stellar system in which it occurs and, if this system is smaller than the Milky Way or the Andromeda nebula, even brighter, as shown in Plate XVI.

The System of Galaxies as a Fair Sample of the Universe

The determination of the distances of the galaxies, or extragalactic nebulae, is the basic problem of observational cosmology. After nearly half a century of intensive research it remains a subject of baffling complexity. Before considering it further, let us briefly discuss other lines of investigation which support the hypothesis that the galaxies are independent stellar systems.

We have seen that the distribution of stars in our own galaxy shows a marked tendency to concentrate towards the main plane of the Milky Way. The extragalactic nebulae as actually observed in the sky show the opposite tendency. In fact, there is a zone of avoidance where there appear to be no external galaxies, coinciding with the region where the stars are thickest. No galaxies are seen within a band, varying from 10° to 40° in width, running along the central region of the Milky Way. Outside this band the numbers actually observed are found to increase as the telescope is directed away from the Milky Way towards the galactic poles. Hence, the apparent distribution of galaxies is very different from that of the stars, a result which can be most naturally explained as an optical effect due to the obscuring influence of a zone of dark absorbing material running round the main belt of the Milky Way.

The existence of such a belt of obscuring matter running round the whole system can be seen very clearly in some other spiral nebulae. A good example is shown in Plate XXIV. It is presumed that there is a very similar zone in the Milky Way, and if we could look at it from outside, from a direction similar to that in which we observe this spiral, its appearance would be similar. When the observed number-counts of the extragalactic nebulae are corrected for the effect of such a zone of obscuration running round our own system, there are no major departures from isotropy, the distribution of the external nebulae being much the same in each direction.

Our relation to the external galaxies must, therefore, be very different from our relation to the stars. The solar system, as we have seen, is markedly eccentric in a large bun-shaped system of stars which rotates with its main plane at right angles to the axis of rotation. With respect to the galaxies, on the other hand, it is more or less centrally situated in a sphere which shows no evidence of any axis of rotation or main plane. In order to study the distribution of galaxies in different directions, instead of relying on the giant reflecting telescopes which pin-point very small areas in the sky, a 48-inch Schmidt telescope which has a much wider angle of vision has been used on Mount Wilson. With its aid a new atlas of the whole sky has been prepared to a higher degree of apparent magnitude than in any previous atlas, but much detailed statistical analysis is needed for astronomers to see whether it confirms the hitherto generally accepted view that the large-scale distribution of the galaxies is isotropic.

On the basis of present observational evidence concerning the extragalactic nebulae the following conclusions can be drawn. Either we are embedded in a super-system of galaxies of such enormous extent that in no direction, even with modern resources, have we any knowledge of the rim, or alternatively, as most astronomers believe at present, the aggregate of galaxies so far observed is a fair sample of the grand system which forms the whole physical universe. In one of his last astronomical lectures Jeans remarked that the history of astronomy is 'a history of receding horizons'. Here at last we seem to glimpse the ultimate horizon.

The Classification of Galaxies

The spiral used to be regarded as the dominant shape of the extragalactic nebulae: indeed, the term 'spiral nebula' came to be used as a general name for all external stellar systems. Recent observations, however, have shown that the spiral form is by no means the unique shape, or even the most common, elliptical galaxes, particularly dwarf ellipticals, being more numerous. The situation is somewhat similar to that observed when comparing the apparently brightest stars and the nearest stars, which are in the main quite different. The spiral galaxies, like the brightest stars, are those which are most readily observed because they are the brightest. The ellipticals tend not to show up so readily in the photographs. Although, as their name implies, they appear elliptical in shape, their three-dimensional forms are probably spheroidal. They differ in many ways from spirals. Not only do they possess no spiral arms, but they also contain very little dust or dark absorbing matter.

Besides spiral and elliptical galaxies, there are a number of amorphous structures which are found to be stellar in composition. Prominent among these irregular galaxies are the Magellanic Clouds, two satellite systems of the Milky Way. (The cores of some spirals are bar-shaped, and G. de Vaucouleurs has argued that the Large Magellanic Cloud may actually be of this form.) All these different shapes – spiral, elliptical and irregular – occur intermingled in space.

The classification of galaxies by shape was originally due to Hubble, who died in 1953. Four years later, Morgan and Mayall approached the problem of classification from another point of view. Although Hubble had been careful to avoid the conclusion that his classification implied a particular evolutionary sequence, most astronomers tended to assume that a normal galaxy first appears as a more or less spherical elliptical which in due course develops into a lenticular form and then *either* becomes a normal spiral with tightly wound arms, designated Sa, and after that develops into the increasingly more open types Sb and Sc, *or* becomes a barred spiral with a corresponding sequence of forms, designated SBa, SBb and SBc. The new classification by Morgan and Mayall depends instead on the degree of central concentration of luminosity. According to a somewhat simplified version of their scheme, the galaxies are grouped into seven classes. In class 1 are placed the irregular galaxies, in some of which occur several irregularly spaced concentrations of hot stars and gas. In class 2, small central condensations can be observed, in subsequent classes, up to 6, the central nuclei become increasingly large and luminous and the spiral arms become more tightly wound and relatively less conspicuous. Finally, in class 7 are placed the various types of ellipticals, in which there are no arms, no evidence of dust or gas, and the light is mainly provided by giant yellow stars. Hubble's distinction between normal and barred spirals is retained, but the main criterion for distinguishing the different classes of galaxy is the spectral type of the inner region, or nucleus.

The Nebular Clusters

Despite the large-scale uniformity of distribution of the extragalactic nebulae in all regions of the sky, the small-scale distribution is anything but uniform. Galaxies tend to congregate in clusters, some in small groups of a few and others in massive systems containing hundreds and in some cases perhaps thousands of members.

The nearest and most conspicuous cluster is seen in the constellation of Virgo. It contains some five hundred members. Another well-known cluster is the Coma cluster which is nearly one hundred million light-years away and contains possibly a thousand or more galaxies in a region of diameter between ten and twenty million light-years.

Despite this clustering tendency – many other clusters have been identified besides these two – the ratio of the distances between neighbouring galaxies to their individual diameters is far smaller than in the case of stars. Whereas, apart from individual members of binary and other systems, the stars are, in general, extremely isolated from one another (for example, the distance from the sun to the nearest star is more than twenty million times the sun's diameter), the average distance between neighbouring galaxies is of the order of only ten times their diameters; and the relation between the average distances and diameters of neighbouring clusters of galaxies is of much the same order of magnitude.

Recent statistical tests appear to have eliminated the possibility that the observed distribution of the galaxies, corrected for the effect of the absorbing band around the Milky Way, may be due to some selective effect of viewing associated with intervening interstellar clouds in our own galactic neighbourhood. Instead, it is thought that nearly all galaxies occur in clusters, the spatial distribution of the cluster-centres being more or less uniform. Hence, it seems that these clusters, rather than the individual galaxies, form the principal units of the physical universe.

The Local Group

Our own galaxy belongs to a small cluster known as the 'local group', which contains nearly twenty members within a region of diameter about 3 million light-years. Our galaxy and M31 (the Andromeda nebula) lie nearly at the opposite ends of a diameter and are the two most prominent members. There is one other spiral in the group, M33, which is much nearer to M31 than to our galaxy. There are ten ellipticals in the group of which six are dwarfs. There may well be a few undiscovered members, especially in the regions hidden by the dust clouds of the Milky Way, and there may be some globular clusters and systems of comparable size that are too remote to belong to our galaxy.

Recently two systems have been observed in the infra-red that are believed to be galaxies lying in regions that are obscured from us optically by the gas and dust in the Milky Way. They have also been studied at radio wavelengths and their angular size determined, thereby confirming that they are of the size of galaxies. It is not yet clear, however, whether these two systems, known as Maffei 1 and Maffei 2 after the discoverer of the former, lie just inside or just outside the local group. They lie close in the sky to IC 342, the fourth largest spiral galaxy known, which is beleived to be outside the local group, and it is possible that the Maffei galaxies are satellites of it. All three are very difficult to observe because of the direction in which they lie. If they form a small group that is passing near our local group, it will be interesting to see what effects this close passage may be producing.

The Discovery of the Two Stellar Populations

Although very bright stars, such as novae and cepheid variables, were identified in some extragalactic nebulae nearly fifty years ago, only in 1943 was it conclusively shown that the ellipitical nebulae and the central regions of spiral nebulae are, like spiral-arms and irregular nebulae, also composed of stars. In that year Walter Baade, using the 100-inch Mount Wilson reflector (the 200-inch had not then been completed), finally succeeded in resolving the central region of the Andromeda nebula and also two satellite elliptical nebulae into their stellar components. As already shown, this resolution has since been effected without difficulty using the 200-inch telescope. Nevertheless, Baade's success in 1943 with the smaller telescope was one of the landmarks of modern astronomy, not simply because he had at last verified a result that had long been expected, but because of the method which he employed and the reason he advanced for its success where all previous attempts had failed.

Baade began from the fact that the brightest stars in spiral arms are main-sequence blue stars of high surface temperature. In order to pick out bright stars in the central core of the Andromeda nebula, he therefore began by using blue-sensitive photographic plates. With the sky free from artificial light, owing to the war-time black-out of the neighbouring towns of Los Angeles and Pasadena, and with particularly good seeing conditions in one of the best regions of the world for astronomical observation, it seemed that with the latest blue-sensitive plates success ought to be within reach. Nevertheless, the general fluid appearance of the central region of the Andromeda nebula remained unchanged. It then occurred to Baade to try, instead of blue-sensitive plates, a new type of red-sensitive plate, despite the fact that no red star seemed likely to be detected at so vast a distance when the much more brilliant blue stars had failed to show up in the photographs. With this red-sensitive plate success was immediate. The central core dissolved into myriads of star images. Similarly the companion elliptical nebulae were also immediately resolved into assemblages of stars. It was thus found that, whereas the brightest stars in the spiral arms are blue super-giants, the brightest stars in the hitherto unresolved galaxies and parts of galaxies are red. Moreover, once the threshold of resolution had been reached, these brights stars appeared in great numbers. Their absolute luminosities are about five magnitudes higher, i.e. about a hundred times fainter, than those of the brightest stars in the spiral arms. They are, however, distinctly brighter than the red giant stars of our own part of the Milky Way, and are similar to those found in the globular star-clusters which surround our galaxy in a vast system of spheroidal shape.

Baade came to the conclusion that, contrary to previous ideas, there must be what he called two distinct types of 'stellar population', which he called Population I and Population II. Population I is found in the arms of spiral nebulae and in irregular nebulae such as the Magellanic Clouds. With the exception of novae, the brightest stars known are very hot blue stars, and these are all Population I stars. The brightest stars of Population II are considerably fainter. Population II is found in the central regions of spiral nebulae, in elliptical nebulae and in globular clusters. The brightest stars occurring in these are red. As we pass to fainter stars, the two populations show a general tendency to merge when plotted in a Hertzsprung–Russell diagram (luminosity plotted against spectral class). Whether there is an actual merging or not in the case of very faint stars is not yet certain.

Elliptical and irregular galaxies thus have complementary structures and stellar populations, elliptical galaxies being Population II systems and irregular galaxies, on the whole, Population I systems. Spiral galaxies are more complex; they combine the properties of both, the outer arms being Population I and the central regions Population II. If the Andromeda nebula, for example, is photographed in infra-red, the spiral form completely vanishes because it is mainly picked out by the very hot blue stars. Instead we see a spheroidal Population II system, dominated by the central core. This suggests that the Andromeda nebula is a vast system of Population II stars on which the spiral arms of Population I have been superimposed. Photometric studies have shown that, although the arms appear conspicuous on photographs and are dominated by very bright blue stars, they contribute less than one-fifth to the total light emitted by that galaxy.

The contrast between the two types of stellar population extends to their rotational properties and also to their association with interstellar dust. Population I systems rotate and hence are relatively flat, particularly spiral arms, whereas Population II systems show much less evidence of rotation and at the same time are relatively spherical. Even more striking is the association of Population I systems with regions which are rich in interstellar dust, Population II systems being associated with regions which are practically transparent. Baade has shown that the presence of interstellar dust may be directly correlated with the occurrence of highly luminous blue stars. He found such stars in small dusty (light-absorbing) regions inside two elliptical companions of the Andromeda nebula, whereas an otherwise similar third companion showed no trace either of dust or of blue stars.

In recent years it has become apparent, in particular following the discovery of systematic spectral differences between globular clusters and groups of stars in the disc, or main plane, of the galaxy, that Baade's simple division of stars into two discrete populations is too crude and must be replaced by a more refined classification. Five divisions, forming a continuous sequence, have been suggested: Halo Population II, which includes the globular clusters; Intermediate Population II; Disc Population, which includes stars of the central core; Intermediate Population I; Extreme Population I, which includes super-giants and cepheid variables.

The Scale of the Universe

Since cepheid variables can be identified only in the nearest extragalactic nebulae, other methods must be found for estimating the distances of more remote galaxies. During the years

1924–9 Hubble developed a step-by-step technique. First he found that the brightest constituent stars (other than novae) were all of about the same absolute magnitude, averaging about 50 000 times that of the sun. This result provided a criterion for estimating the distances of those nebulae in which individual stars could be detected. Furthermore, most of these nebulae have luminosities between one-half and twice the average, and this result was then used statistically to estimate the distances of still more remote clusters of nebulae. Thus, a provisional distance-scale for the physical universe was built up. It was a rough-and-ready method and was based on a number of bold hypotheses. Nevertheless, Hubble's scale stood the test of time for rather more than twenty years.

Then, quite unexpectedly, in 1952 Baade announced that Hubble's scale would have to be drastically revised – not merely at the uncertain far end, but at the near end, too. Instead of assigning a distance of about 750 000 light-years to the Andromeda nebula, we would now have to regard it as being at least twice as far away. Indeed, the most recent figure that has been suggested is about 2 200 000 light-years. This corresponds to a scale revision factor of rather more than two.

Why was this startling revision made? A number of factors led independently to the same result and together they made an overwhelming case. On the old distance-scale, observers had been led to expect that, with the aid of the 200-inch telescope which came into operation in 1949, the globular clusters which surround the Andromeda nebula in much the same way as our own Milky Way is surrounded should be readily resolved, so that their brightest constituent stars (as well as comparable stars in the main body of the nebula) could be studied with ease.

The 200-inch telescope can photograph objects down to an apparent magnitude of 22·5. If the Andromeda nebula were at the distance that had been previously estimated, about three-quarters of a million light-years, then the short-period cluster-type variables in that nebula, and in the globular clusters associated with it, should be of apparent magnitude about 21·8. Consequently, many of these variables (RR Lyrae variables) which are all of about the same absolute magnitude should have been readily detected by the 200-inch telescope. But none could be found on the photographs. With an exposure time of thirty minutes that ought to have revealed these stars quite clearly, there appeared only the brightest Population II stars which are known to be about one and half magnitudes more luminous than the cluster-type variables. Therefore, it looked as if the distance-scale for the Andromeda nebula was wrong by a factor of about two, corresponding to a difference on the magnitude-scale of one and a half. In other words, the Andromeda nebula must be about twice as far away as had previously been thought.

This was not the only line of argument leading to the revision of the extragalactic distance-scale. There were other inconsistencies associated with the old scale. For example, the globular clusters which surround the Andromeda nebula appeared to be systematically smaller than those surrounding our own galaxy. On the average they seemed to be of only about half the diameter. The discrepancy would disappear if they were twice as far away as previously supposed.

Moreover, the average absolute magnitude at maximum brightness of novae in the Milky Way is about −7·4, whereas that of novae observed in the Andromeda nebula was only about −5·7. Here again the discrepancy in magnitude is very roughly one and a half. So once more the apparent anomaly could be removed by adopting a correction factor for distance of about two.

The original mistake is thought to have arisen in the determination of the absolute scale of distance, not for the short-period RR Lyrae variables, but for the longer period classical cepheids, those for which Miss Leavitt had obtained the period-liminosity law. As mentioned earlier, the conversion of her scale of relative distances into an absolute scale was based on some independent direct determination of the distance of one cepheid and evidently a mistake had been made. The classical cepheid variables for which the distance-scale was wrong are Population I stars, whereas the RR Lyrae short-period variables are Population II stars. Indeed, Baade's discovery of the new distance-scale was intimately related to his previous discovery of the two stellar populations.

On the old scale it had appeared that our own galaxy was

larger than any other stellar system, a result which had always been regarded as somewhat anomalous. According to the revised calculations, since the Andromeda nebula has been found to be twice as far away as previously thought, its diameter is about twice as great and hence it is larger than the Milky Way. The dimensions of the latter were completely unaffected by the new scale because they had been obtained from the study of RR Lyrae variables in the surrounding globular clusters, and the scale of distance of these remained unaltered. Thus, our own system, although one of the largest of the galaxies, is in fact not the largest of all. According to the latest estimate, based on studying the radio waves from the Andromeda nebula, it is claimed that this galaxy has a main diameter of the order of 250 000 light-years compared with about 100 000 light-years in the case of the Milky Way, but this figure for the Andromeda nebula may not be final.

Baade's revision of Hubble's distance-scale was followed in the autumn of 1957 by a further revision due to Sandage. This concerned the second step in Hubble's technique, his brightest star criterion. Sandage showed that objects in the near-by Virgo cluster of galaxies which Hubble believed to be highly luminous stars are in fact regions of glowing hydrogen gas of intrinsic luminosity about two magnitudes brighter than that postulated by Hubble on the assumption that they were stars. Consequently, the distances formerly assigned by Hubble to all galaxies beyond those in which cepheid variables can be detected with our most powerful telescopes must now be multiplied by a factor between 5 and 10. The observable universe is therefore even more extensive than was thought following Baade's revision of its scale. We now believe that with the most powerful optical telescope, the 200-inch reflector on Mount Palomar, galaxies can be detected up to distances of perhaps 5 000 million light-years. We see galaxies right up to the present limit of our optical vision. So far there is no sign of any boundary. Indeed, it is thought that radio waves are being received from even more distant objects than any which can be detected visually.

The Masses of the Galaxies

We have already seen that from study of the rotation of our galaxy it has been deduced that its total mass is between one and two hundred thousand million times that of the sun. This has been calculated on the basis of Newton's theory of gravitation. Similarly, masses of other near-by galaxies have been estimated by measuring their rotation and invoking Kepler's third law. In this way it has been found that the mass of the Andromeda nebula is of the order of 300 000 million times that of the sun. Of particular interest are the properties of the bright nucleus of this galaxy, for in a region of diameter 25 light-years are crowded together stars of total mass about 13 million times that of the sun, giving a mean density of about ten thousand times that in our neighbourhood.

Another method of measuring the masses of galaxies depends on the fact that galaxies often occur in pairs, and is similar to that of determining the masses of binary stars. The main difference is that, whereas stars take from a few days to a few hundred years to go around each other, galaxies take a few hundred million years. Consequently, we have to rely on Doppler-shift measurements that refer to only a part of the full orbital speed, depending on the unknown angle of the motion to our line of sight. However, if we assume that the relative orbits are oriented completely at random, we can expect that the average of a few dozen pairs will give a fairly reliable value for the mean mass of a galaxy. The results obtained give an average value for the masses of spirals and irregulars of 40 000 million times the mass of the sun and for ellipticals of 700 000 million times the mass of the sun. It would seem that our galaxy and the Andromeda nebula are relatively large and massive spirals. On the other hand, although many ellipticals are dwarf galaxies, the range of size of this type is much greater than that of spirals. Among the giant ellipticals is M87 which has a mass some twenty times that of the Milky Way. Dwarf ellipticals of masses a few thousand million times that of the sun appear to be the commonest type of galaxy.

A third way of measuring the average masses of galaxies applies to those that exist as members of a cluster. If we assume that a cluster is a stable group that is not losing members to outer space and we can determine the relative velocities of the individual galaxies composing it, we may invoke a general law of dynamics, known as the Virial Theorem,* to calculate the total mass of the cluster. On dividing this by the total number of galaxies in the cluster, we obtain the average mass of a galaxy. But here we find a discrepancy, since the average mass of a galaxy determined in this way comes out to be between ten and a hundred times greater than that obtained by the other methods.

Two possible explanations of this result have been considered. The Virial Theorem may not apply because the clusters may be expanding and breaking up, but if so how is it that there are so many clusters left? Alternatively, there may be many invisible galaxies in the clusters and large quantities of intergalactic matter, amounting to ten or more times the total mass of the galaxies. At present this question is still unresolved, but the latest evidence is against there being sufficient intergalactic matter to make a large cluster such as the Coma cluster gravitationally stable.

Intergalactic Matter and the Intergalactic Wind

Once the existence of interstellar matter had been established it was natural to ask whether there is also diffuse matter between the galaxies. Hubble, one of the greatest of all observers in this field, came to the conclusion that extragalactic space is remarkably transparent, but in recent years evidence has accumulated that in between the different galaxies of some clusters there may be diffuse matter. At least one investigator has reported evidence for the absorption of light from remote clusters lying behind nearer clusters; this might be due to intergalactic dust in the latter.

In 1969, Oort reported that observations with a radio telescope equipped with a high sensitivity 21-centimetre line receiver indicated the existence of a considerable flux of gas flowing into the Milky Way from outer space. From the observed velocities

*This asserts that $2E + V = 0$, where E is the time-average of the kinetic energy and V is the time-average of the potential energy, for a steady state.

and densities, he concluded that over the whole galaxy the inflow corresponds to about 3 solar masses a year, so that in a thousand million years the total mass of the galaxy increases by nearly 2 per cent. Oort calculated that the flow of this 'inter-galactic wind' corresponds to a density just outside our galaxy of about 3×10^{-28} grammes per cubic centimetre, but he pointed out that this must be higher than the actual intergalactic density because the gas has been concentrated towards and accelerated by the galaxy and other near-by systems. In the local group as a whole, Oort believes that there may be three times as much intergalactic matter as there is matter concentrated in the galaxies.

Newton's Universe

Despite recent spectacular advances in our knowledge of very distant objects, there is no general agreement concerning the full extent of the physical universe, and in particular on whether it is finite or infinite. Conflicting views on this problem can be traced back to antiquity. Aristotle thought that the physical universe must be finite, whereas the ancient atomists, for example Lucretius, believed that there were atoms throughout the whole of infinite space. Lucretius says in his poem, 'Space is without end or limit and spreads out immeasurably in all directions alike.' During the mediaeval period thinkers tended to follow Aristotle in regarding the universe as finite. The first astronomer to take the step of likening the stars to our own sun and scattering them throughout infinite space was an Englishman, Thomas Digges, about 1576. This view was enthusiastically adopted by the ill-fated Giordano Bruno, who was burned at the stake for heresy in 1600.

Following the establishment of his theory of universal gravitation, Newton was led, on dynamical grounds, to embrace the idea of an infinite universe. He argued that a finite universe in infinite space would tend to concentrate in one massive lump under its own gravitational attraction. In one of his letters to Bentley in 1692 he wrote, 'But if the matter was evenly disposed throughout an infinite space, it could never convene into one mass; but some of it would convene into one mass and some into another, so as to make an infinite number of great masses scattered at great distances from one to another throughout all that infinite space.' Nevertheless, he fully realized that this argument presupposed that his law of gravitation was universal. Not for another hundred years, until the researches of Herschel and W. Struve, was it definitely known that the law could be extended beyond the solar system.

In 1895 a peculiar difficulty in Newton's argument was pointed out by the German astronomer H. Seeliger. He began, like Newton, by assuming that matter is more or less uniformly distributed throughout the whole of infinite Euclidean space. He then considered all the matter within a sphere of radius R, its mass being proportional to its volume and hence to the cube of R. At any point on the surface of this sphere the gravitational attraction towards its centre will be proportional to the mass and inversely proportional to the square of R, the distance from the centre. Hence, the attraction will be directly proportional to R. If the whole universe is infinite, as Newton supposed, we can consider it as a sphere of infinite radius, but in that case there will be an infinitely intense gravitational field at points infinitely far from its centre. This centre, however, can be chosen to be anywhere we please. It seemed to Seeliger that this argument led to absurdity, and he therefore suggested that, to avoid having infinite gravitational intensity everywhere in space, Newton's law must be modified at great distances. He proposed a new term in this law, which would be effective only on the cosmic scale.

Einstein's Universe

Twenty years later, a much more profound modification of Newton's law of gravitation was made by Einstein in his general theory of relativity, which explained all that the Newtonian theory could account for and several other phenomena in addition. Shortly afterwards, in 1917, Einstein applied his new theory of gravitation to the structure of the whole physical universe. He was so impressed by Seeliger's argument and by other arguments of a similar nature that he took the drastic step of 'abolishing infinity'. Instead, he maintained that the universe as a whole was finite and unbounded. Its geometry, therefore, could not be the ordinary Euclidean geometry of infinite space but another type associated with a finite, unbounded space. Such a space may be regarded as a three-dimensional analogue of the two-dimensional surface of a sphere. Just as we can travel continuously over the finite surface of the earth without coming to any point where that surface can be said to end, so in a finite unbounded universe there would be no outer boundary of space.

In Einstein's theory, as in all previous theories concerning the structure of the universe, the universe *as a whole* is static. This means that all celestial motions are thought to be negligible compared with the speed of light which plays an important role in Einstein's theory.

In 1930, however, Eddington discovered that Einstein's universe was unstable and would tend either to contract or to expand. Nevertheless, Eddington regarded the finite Einstein universe as the original state of the actual universe and believed that its physical properties determine the laws of nature which control the world as we know it.

Einstein did not assign any precise size to his world-model, but showed that its mass was directly proportional to its radius. Eddington went further and determined a definite value for each. By an ingenious but extremely difficult argument, based on a peculiar conception of the nature of physical measurement, he claimed to obtain a precise value for the number of nucleons and electrons in the world. This number was of the order of 10^{79}, the corresponding mass being about 10^{55} grammes. This would mean that there must be enough material in the universe to form roughly a 100 000 million galaxies, each containing 100 000 million stars of average mass equal to that of the sun. The radius of an Einstein universe of this size would be of the order of 1 000 million light-years, but owing to the instability of the system this could only be its initial value. Considerations of this kind show that in analysing the structure of the actual universe we must examine not merely its spatial properties but also the problems of its origin and evolution.

CHAPTER 12

The Evolution of the Universe

The Background Brightness of the Night Sky

We have seen how Newton was led by a dynamical argument based on his hypothesis of universal gravitation to the conclusion that matter must be more or less uniformly distributed throughout the whole of infinite space. Furthermore, we have seen how this conclusion led Seeliger some two hundred years later, at the end of the nineteenth century, to propose a modification in the law of gravitation so as to avoid the paradox of infinite gravitational intensity. Early in the nineteenth century, however, another German astronomer, H. W. M. Olbers, had already formulated a paradox concerning the intensity of radiation in Newton's universe. He pointed out that an infinite universe of stars uniformly distributed, as envisaged by Newton, would result in infinite sky brightness because, although the apparent brightness of an average star at distance R will vary inversely as the square of R, the number at that distance will be roughly proportional to this square, since the surface area of a sphere is proportional to the square of its radius. Hence, if the universe is infinite, it should be observed as infinitely bright.

Although Olbers's argument can be modified by allowing for the interventions of clouds of obscuring matter and by making assumptions about the size and the age of the universe, its essential basis is the postulate that both the distribution and the intrinsic properties of radiation-sources are uniform throughout the universe. The argument would collapse if this assumption were modified in some way by introducing a time-effect or a space-effect. For example, a lower intensity might be attributed to a distant light-source at the time when the observed radiation was emitted, than to a similar source observed within our own neighbourhood. Alternatively, the energy of each quantum of radiation from a distant source might be less than that of a corresponding quantum of radiation from a near-by source. Since the energy of radiation is inversely proportional to its wavelength, this reduction would be associated with an apparent increase in the wavelength throughout the whole spectrum which would be shifted to the red. Such a phenomenon could arise if the distant source were not at rest but were receding from us.

The Extragalactic Red-Shifts

The pioneer of extragalactic spectroscopy was V. M. Slipher of the Lowell Observatory, who in 1912 obtained the first spectrum of the great nebula in Andromeda. This spectrum was shifted towards the blue. When such a phenomenon had been observed in stellar spectra it had usually been attributed to the Doppler effect associated with motion in the line of sight, a blue shift indicating that the source was approaching, a red one that it was receding. Slipher, therefore, assigned to the Andromeda nebula a velocity of approach of about 125 miles, or 200 kilometres, per second. Although high, this was not a fantastic speed to encounter in astronomy; average speeds for different spectral types of star attain nearly 100 kilometres per second. However, by 1917 Slipher had obtained spectra of fifteen spiral galaxies, all but two of which were displaced to the red by amounts corresponding to speeds of recession averaging 400 miles, or 640 kilometres, per second.

Although these speeds were much greater than any previously assigned to the stars in our own galaxy, the standard procedure of attributing such shifts to the Doppler effect associated with motion in the line of sight was not immediately called in question. Moreover, these speeds were all less than 1 per cent of the velocity of light, and consequently might not have been expected to lead to any serious conflict with the postulate explicitly adopted by Einstein in 1917 when constructing his finite world-model, that the universe in the large is effectively static. Astronomers were, however, perplexed not so much by the magnitude of these velocities as by their orientation, for nearly all were directed outwards, which was difficult to reconcile with the basic hypothesis of the static universe that the proper motions of the galaxies were purely random. A systematic motion of recession, whatever its magnitude, would appear to contradict this hypothesis.

To resolve the question at issue it became imperative to obtain spectrograms of more distant galaxies with the most powerful light-gathering instruments available. Fortunately, the 100-inch reflector came into service on Mount Wilson early in the 1920s. The task was long and arduous; exposures lasting several nights were necessary and the greatest skill was required to keep the source steadily fixed over the slit of the spectrograph night after night, particularly when, even with this powerful telescope, it remained invisible to the human eye. The resulting picture was often little more than one tenth of an inch long and one thirtieth of an inch wide. On such barely perceptible data a new conception of the physical universe was to arise.

Despite the difficulty of obtaining spectra of these remote objects there was no doubt of the reality of their red-shifts. These spectra are dominated by the so-called H and K absorption lines of calcium which are among the most prominent lines in the spectra of stars like the sun. We know that in our own Milky Way the sun is an average type of star, and the extragalactic nebulae, particularly the spirals, presumably contain large numbers of similar stars. The red-shifts are measured by comparing the positions of the H and K lines of calcium in the extragalactic spectra with their position in the solar spectrum. If the shifts are due to recessional motion then the speeds can be immediately deduced from these measurements.

By 1929 Hubble had formulated his brightest-star criterion for the distances of those galaxies which lay beyond the region in which cepheid variables could be detected with the aid of the 100-inch telescope. With this method he assigned distances to galaxies outside the local group up to six million light-years. (On Sandage's revised distance-scale this would correspond to a distance of forty or more million light-years.) For all galaxies within this range, but outside the local group (which includes the Andromeda nebula), there appeared to be a definite linear correlation between speed of recession and distance so that, if one galaxy were twice as far away as another, its spectrum would be shifted approximately twice as much. Later, Hubble found that this relationship also applied to more remote nebulae whose distances he had estimated on the hypothesis that there is a definite upper limit to the absolute magnitude of a galaxy.

After 1929 Milton Humason exploited the potentialities of the 100-inch reflector to photograph the spectra of increasingly fainter and more distant galaxies. Within a few years he obtained a remarkable series of photographs indicating a progressive reddening of spectra with increasing apparent magnitude and distance (see Plate XXV). At the limits of observation with the Mount Wilson instrument he recorded red-shifts corresponding to velocities of recession up to 40 000 kilometres per second, i.e. velocities up to nearly one seventh that of light.

After 1949, when the 200-inch telescope on Mount Palomar came into service, Humason succeeded in photographing the spectra of even more distant galaxies, including those of two members of the remote Hydra cluster. One of these is illustrated at the bottom of Plate XXV. The velocity of recession corresponding to the red-shift shown in this spectrum is about one fifth of the velocity of light. Several years later, in 1956, with the aid of special photo-electric equipment attached to the new telescope. Baum obtained a red-shift corresponding to a recessional velocity of about two fifths of the speed of light. In 1960 an even greater shift was determined by Minkowski corresponding to nearly half the velocity of light. Far larger red-shifts have since been determined following the discovery of quasars (see p. 278), corresponding to velocities up to early nine tenths that of light.

The Expanding Universe

The initial hypothesis that the extragalactic red-shifts were due to the Doppler effect associated with motion in the line of sight was not disputed until the early 1930s when the predominance of abnormally high velocities became apparent. It was clear that a new phenomenon of nature had been discovered and alternative explanations were explored. For example, it was suggested that the shifts might be due to intense gravitational fields associated with distant galaxies, Einstein's general theory of relativity having predicted that strong gravitational fields could give rise to red-shifts. Alternatively, the shifts might be the result of some hitherto unsuspected ageing of light in transmission over the vast distances of internebular space by which it automatically loses energy.

For one reason and another none of these alternative *ad hoc* hypotheses has received anything like the degree of support given to the original Doppler interpretation and the associated recessional hypothesis. So far there is no observational evidence which can be regarded as directly conflicting with it. An essential feature of the Doppler effect is that, for a given source of light the ratio of the observed wavelength of any line in its spectrum, for example the H or the K line of calcium, to the corresponding wavelength observed in the laboratory should be the same for all lines. It is not easy to test this feature in the case of extragalactic spectra, but its occurrence in the spectra of some twenty galaxies has been verified to within a probable error of about 5 per cent over a considerable range of wavelengths. In the case of one galaxy exceptional accuracy was possible and the feature was established to within one per cent.

Much of the objection to the Doppler interpretation of the extragalactic red-shifts has been basically psychological: the velocities obtained were so 'fantastic' that it seemed difficult to believe in their reality. However, they cease to be so incredible if, instead of expressing them in kilometres, or miles, per second or as fractions of the velocity of light, they are considered in relation to the diameters of the galaxies. The earth in its orbital motion around the sun takes ten minutes to pass through a distance equal to its diameter. The sun travelling at about 250 kilometres per second around the centre of the galaxy describes a distance equal to its diameter in about an hour and a half. How long would a distant galaxy, comparable to our own Milky Way in size and receding with one fifth of the velocity of light, require in order to pass through a distance equal to its main diameter?

The time is of the order of half a million years. This way of looking at the problem does not, of course, prove that the Doppler interpretation is the correct one, but it diminishes the psychological objection.

It may seem curious that the most prominent galaxy in the sky, the great nebula in Andromeda, is an exception to the general rule and shows a blue-shift in its spectrum. This peculiar shift is now believed to be primarily due to the fact that at the present time the rotation of the galaxy is swinging the solar system round in the direction of this nebula. More generally, Hubble's law relating distance and spectral shift does not hold within the local group of galaxies, presumably because the tendency to individual recessional motion is masked by proper motions within the local gravitational field of the group. It is now thought that Hubble's law applies to clusters of galaxies, and if we regard these clusters as the principal units of the physical universe, then the universe as a whole cannot be in a steady state, as both Newton in 1692 and Einstein in 1917 assumed, but instead must be expanding in all directions.

Evolutionary Theories of the Universe

There are at present two main difficulties in studying the expansion of the universe. The first concerns the behaviour of the redshifts in time, and the second relates to the establishment of a reliable scale of distance.

We have no direct evidence bearing on the variation of redshifts in time because the phenomenon has only been observed for about half a century, a minute fraction of the time taken by light to travel to us from any extragalactic nebula. Nevertheless, Hubble's law relating shifts or velocities to the corresponding distances is a spatio-temporal law. For, when we deduce from observations that a particular galaxy is 500 million light-years away, what we mean is that it was at that distance 500 million years ago, whereas all galaxies within the local group are seen as they were not more than 2 million years ago. The simplest hypothesis to adopt for the temporal behaviour of the spectral shift of any galaxy, as it would appear to a sufficiently long-lived observer, is to assume that it does not change with lapse of time. Then, on the recessional interpretation, it would follow that the clusters are all receding from each other with more or less uniform velocities.

This particular hypothesis was adopted by Milne in 1932. He pointed out that any system of bodies moving uniformly in all directions would in the course of time tend to become an expanding system in which the fastest would have receded the farthest, the respective distances and speeds tending ultimately to obey the law $r = vt$, where r denotes distance, v denotes speed and t denotes time. This law is of the same form as Hubble's empirical law, but provides the additional information that the 'constant' of proportionality in the correlation of speed and distance is a measure of the time that has elapsed since the whole system was in a state of maximum concentration and density.

Guided by this simple way of looking at the situation, Milne then constructed, with the aid of the special theory of relativity, a theoretical world-model in which the galaxies were all in uniform recessional motion from each other. He came to the conclusion that, if this model satisfactorily represented the main features of the actual universe, then all the galaxies must have been compressed together in a comparatively small volume a finite number of years ago, that number of years being given by the empirical value of t in Hubble's law.

The principal uncertainty in the evaluation of t is due to the second difficulty mentioned above, the determination of a reliable scale of distance. When Milne first put forward his theory in 1932, the scale of distance already constructed by Hubble led to the result that the value of t was about 2 000 million years. Milne's theory, therefore, implied that this was the age of the universe since expansion first began.

The Milne universe depends on the assumption that, since each galaxy is assumed to be at a centre of symmetry, the net gravitational pull on it is zero and therefore it is possible to regard the recessional motion of all galaxies as uniform. On the other hand, galaxies in world-models constructed in accordance with Einstein's general theory of relativity are subject to a slowing-down of their recessional motion under the influence of gravitation. The simplest of such expanding world-models, known as the Einstein–de Sitter universe, expands from a point-like origin, as does Milne's, but since its expansion was faster in the past than now, its age is less than that of the Milne universe, being $2t/3$, where t is given by Hubble's empirical law as before.

An alternative theory, advocated by Eddington and Lemaître, was based on the idea that Einstein's static universe was in a state of delicately balanced equilibrium between two opposing forces, namely, the binding force of world-gravitation and a disrupting force known as cosmical repulsion, effective only on a very large scale and quite insignificant on the scale of the solar system. As mentioned in the previous chapter, Eddington discovered in 1930 that this universe was unstable. A slight radial disturbance would cause it either to expand or contract so that whatever happened it would not return to its original steady state. Eddington assumed that such a disturbance had occurred and that, in fact, the universe was expanding. On this view, expansion must have been slower in the past than now. The age of the universe according to this theory was much greater than according to Milne's.

The Steady-State Theory of the Universe

In 1948 a radically different interpretation of the observed facts was put forward by H. Bondi and T. Gold, and shortly afterwards by F. Hoyle. According to this hypothesis there is no expansion of the universe, only recessional motion of individual galaxies and clusters. The universe as a whole is in a steady state and was never more compressed than now. There is neither an expansion of finite curved space carrying the galaxies with it, as Eddington believed, nor an expansion of the whole system of galaxies into an infinitely extended space, like a gas dispersing into a vacuum, as suggested by Milne. Instead, according to this new theory, both space as a whole and the system of galaxies as a whole show no temporal evolution whatsoever. Only galaxies and stars pass through the successive stages of an evolutionary history, not the whole universe. As old galaxies and clusters of galaxies stream away from each other, a universal steady state is maintained by the continual and ubiquitous creation of new galaxies to fill up the gaps in space that would otherwise appear as the older galaxies drift apart.

At first the new hypothesis was widely acclaimed. It had several attractive features. From the early nineteen-thirties many men of science had sought to escape from the obvious implications of the successive discoveries of Slipher, Hubble and Humason. The usual line of escape was to seek some alternative to the Doppler interpretation of the red-shifts. The new line of escape was more ingenious. The Doppler interpretation was accepted, but its apparently inescapable consequence, that the universe as a whole cannot be in a steady state, was unhesitatingly rejected. In order to keep the system going eternally without any overall change, the principle of the conservation of matter on which, for example, modern chemistry was founded by Lavoisier had to be abandoned. Instead, it was suggested that new matter in the form of neutral atoms of hydrogen was continually appearing *out of nothing* spontaneously and more or less ubiquitously in space, although the calculated rate of creation was so small as to be far below any possibility of direct observation.

This theory, at least in its original form, encounters the following difficulty. An old cluster of galaxies – and in due course any cluster, in particular our local group, must age – would acquire by the continual creation of new matter in and around it, an exceedingly powerful local gravitational field which would more than counterbalance the effect of expansion. (It has already been mentioned that there is no evidence of expansion within the local group.) Moreover, this cluster would become ever vaster in extent, without limit. Since the theory presupposes that all regions of the universe are equivalent, this situation should occur statistically everywhere. Furthermore, as it is postulated that there is no evolution of the universe, what will happen in the future must already have occurred in the past. Hence, it follows that there could be no expansion anywhere at any time.

Despite these conceptual difficulties, there were certain observationally significant features of the theory which attracted widespread attention. In particular, owing to there being no world-evolution, the distribution of different types of galaxy in different stages of evolution should be purely random. Thus, distant galaxies, observed as they were hundreds of millions of years ago, should be indistinguishable statistically from those observed nearby as they were only a few million years ago, and indeed the different types of galaxies, irregular, spiral and elliptical, which may represent different stages of evolution, do appear to be all intermingled in the clusters.

The Time-Scale of the Universe

Another observationally significant feature of the steady-state theory concerns the time-scale of the universe. It was argued that evolutionary theories of the expanding universe allowed insufficient time for stellar evolution and for the past history of the earth. Strictly speaking, such an argument could only be legitimately advanced against some of these theories. For, in his last paper to the Royal Astronomical Society in 1944, Eddington calculated that if the expanding universe had originated, as he believed, in an unstable Einstein universe, its present age would be of the order of 90 000 million years. On the other hand, it is true that by 1948 the apparent inadequacy of time-scale was a serious embarrassment for those who supported any theory of the expanding universe such as Milne's, which involved the interpretation of Hubble's constant as a measure of the present age of the universe. The then accepted value of this constant was such that the age came out to be about 2 000 million years according to Milne and about 1 300 million years according to Einstein–de Sitter. But a careful analysis, two years previously by Holmes, of the radioactive minerals in the earth's crust had led him to conclude that the most probable age of the earth's surface was about 3 350 million years. Despite the uncertainty of the data, it was thought unlikely that the age of the earth could be much less. Consequently, astronomers were confronted with the anomaly that the age of the universe on the simplest expansion theories was less than the age of the earth!

According to the steady-state theory, no such difficulty could arise since the universe as a whole was presumed to have an infinite past. At one time it was even claimed that only the steady-state theory could cope with the facts. However, one of the immediate consequences of the new distance-scale for the extragalactic nebulae, announced by Baade in 1952, was that the empirical value of t in Hubble's law, $r = vt$, had to be doubled since r was doubled and v, being deduced directly from the red-shift, was unaltered. Hence, the age of the uniformly expanding universe, which is equal to t, had to be increased from about 2 000 to about 4 000 million years. According to the latest revision of the distance-scale due to Sandage the age of the expanding universe has been extended still further. The rate of increase of velocity with distance in Hubble's law, according to observational results announced by Sandage in 1971, is about 50 kilometres per second per million parsecs. The corresponding value of t is about 18 000 million years. It is possible that this is a rough estimate of the age of the universe if it has expanded at a uniform rate. If, however, the expansion has been continually slowing down, as in the Einstein–de Sitter universe, then the age of the universe would be nearer 12 000 million years.

The recent dramatic revisions of the distance-scale, and hence of the time-scale of evolutionary world-models, have removed the former anomaly concerning the relative age of the earth and lead to no new difficulties concerning the ages of other bodies as compared with that of the universe. For example, the sun is now estimated to be some 5 000 million years old, and the earth 4 600 million years. Estimates for grouping of stars which range from binaries to vast clusters, are based on consideration of the external influences which tend to disrupt them. It has been calculated that the disturbing effect of the galaxy as a whole on close binaries, for example, has not been acting for more than 10 000 million years at most. The estimated ages of open star-clusters suggest that this figure of 10 000 million years is an upper limit and can probably be reduced to one-half or less.

Globular star-clusters, which are Population II systems, show every sign of great age. They are highly compact (see Plate XXII) and have life-expectancies of more than 10 000 million years. Moreover, they contain no very hot highly luminous stars which burn up their energy comparatively rapidly. Theoretical considerations based on the actual range of stars found in

271

globular clusters lead to the conclusion that all the constituent stars are of about the same age, which is of the order of 10 000 million years.

Turning from clusters of stars to clusters of galaxies, calculations made of the time for which they can be expected to hold together under their gravitational attractions lead to life-expectancies ranging up to about 10 000 million years. These calculations are based on measurements of their internal motions, and these are deduced by comparison of the respective red-shifts of their component galaxies.

Another line of enquiry pointing to a comparable result concerns the natural radioactive elements. Their very existence at the present time as elements which spontaneously decay implies that they have not always existed in their present form. The observed uranium-lead abundance ratio (see Chapter 2) and the half-lives of uranium and thorium are consistent with each other and with the age attributed to the universe. If one assumes that radioactive isotopes were equally abundant when formed, then from their present observed relative abundances one may calculate the epoch of formation. The elements that have half-lives of the order of 10 000 million years, such as uranium-238 and thorium-232, are still abundant on the earth, whereas isotopes of shorter half-lives, such as uranium-235, are much rarer. Calculations show that uranium-235 and uranium-238 would have been equally abundant about 6 500 million years ago.

Thus all the available evidence points to the conclusion that, contrary to what was thought in 1948, the theory of the expanding universe leads to no serious time anomalies.

The Distribution of Radio Galaxies

Crucial evidence in favour of evolutionary theories of the universe and against the steady-state theory (and any other non-evolutionary theory) has been advanced by Sir Martin Ryle and his colleagues at the Mullard Radio Astronomy Observatory of the Cavendish Laboratory, Cambridge. Observing the sky with radio telescopes operating at metre wavelengths, they have found that, besides a continuous background mainly due to the Milky

Way, there are thousands of compact sources, a few minutes of arc in extent or less. More than 8 000 are now known, but only a few hundred have so far been studied in detail. A few lie within our galaxy and are believed to be the remnants of supernovae explosions, the best known being the Crab nebula (see p. 222). Many of the others are associated with faint galaxies and some lie beyond the range of the most powerful optical telescopes. (Owing to the comparatively low variation of the radio emission with wavelength, the red-shift has a much smaller effect on the intensity of a radio-source than on that of a light-source.) From their distances we deduce that the radio emission from these remote sources is very great, in some cases being more than a million times that from a normal spiral galaxy, such as our own or the Andromeda nebula. These powerful sources are known as *radio galaxies*.

Because they lie at great distances, radio galaxies should provide information about regions of the universe as they were thousands of millions of years ago, at epochs earlier than can be reached with optical telescopes. According to evolutionary theories, the universe as a whole is expanding and in the past galaxies were more closely packed together than they are now. If, however, the universe is in a steady state, conditions in the past were much the same as now, and the number of objects of a particular kind in a given volume of space, provided it is not too small, should, on the average, be the same at all epochs. The most exciting result of successive Cambridge surveys of radio sources – the modern counterpart of William Herschel's famous 'sweeps' of the heavens – has been the confirmation of the main result of the first survey, completed in 1955: that the number of extragalactic radio sources increases more rapidly with distance than if they were distributed uniformly in space, and hence in time.

The fact that the distribution of sources is *isotropic*, i.e. much the same in each direction, is a strong argument for the view that they are at cosmological distances. Otherwise, if they were in our galaxy there would be such a concentration of sources in our part of the Milky Way that it would be impossible to reconcile the resulting conception of our galaxy with the observed radio emission of the Andromeda nebula, which we have every reason to believe is a similar, albeit a little larger, stellar system. On the other hand, the lack of *homogeneity* in the spatial distribution in depth of radio sources is a powerful argument against the steady-state theory and in favour of the evolutionary class of world-models.

The actual distribution in depth of radio galaxies, inferred from the determination of the numbers of sources in different ranges of flux density, initially increases much faster than expected for a uniform population, indicating that at earlier epochs the number, or else the intrinsic power, of radio sources was greater than it is now, but beyond distances corresponding to red-shifts * of about 2·5 (about 8 000 million light-years) there appears to be a sudden steep falling-off in the number of sources. The existence of this cut-off is confirmed by study of the total emission from all extragalactic sources, since the contribution from sources that have been studied individually is about half the total and most of the rest must be assigned to sources of lower power at distances corresponding to lower red-shifts.

As Ryle himself has pointed out, these observations provide powerful additional evidence against any explanation of source-counts in terms of a local non-uniformity in the distribution of radio galaxies. The observed isotropy implies that we would have to be centrally placed in such a system. Since the observed sources account for most of the extragalactic background emission, there can be few other similar systems within a distance corresponding to a red-shift of unity (about 6 000 million light-years). In Ryle's opinion, 'We can only accept a local origin if we suppose that we are situated in a specially favoured place in the Universe – a situation which has been distasteful to astronomers since the time of Copernicus.'

Exploding Galaxies

The first identification of a discrete radio source with an optically visible object was made in 1951 by W. Baade and R. Minkowski.

*Red-shifts are measured by z, where $1 + z$ is the ratio of the wavelength of the radiation received by us to that of the corresponding emission.

They found that the Cygnus A radio source, originally discovered by J. S. Hey in 1946, coincided with a system that we now consider to be some 700 million light-years away. It was a strange object that looked like two galaxies in collision (see Plate XXVI). Shortly afterwards the radio sources known as Centaurus A and Virgo A were found to coincide with two peculiar giant galaxies. The former was also interpreted as being two galaxies in collision – an elliptical and a spiral – whereas the latter, believed to be the most massive member of the Virgo cluster, is a giant ellipse from which emerges a brilliant jet of luminous matter several thousand light-years long (see Plate XXVII).

The idea that the main extragalactic radio sources were produced by the interaction of the gaseous components of colliding galaxies was contested by V. A. Ambartsumian, who argued instead that the radio sources in question should be regarded as huge systems in the course of separation into independent galaxies. The decisive argument for rejecting the collision hypothesis came after it was generally agreed that the mechanism by which radio radiation is generated by radio galaxies is the synchrotron process produced by the interaction of electrons moving at speeds close to that of light in a magnetic field. For, on the collision hypothesis, it was found that in the case of two massive galaxies the total kinetic energy involved would have to be converted *entirely* into the energy of relativistic electrons, and no method is known whereby this could happen. Indeed, the maximum energy of two colliding galaxies can hardly exceed 10^{52} joules, whereas the total radio energy output of the most intense radio galaxies is believed to be in the region of 10^{54} to 10^{55} joules. This is equivalent to the total nuclear energy output of between 10^8 and 10^9 stars like the sun; but, whereas such a star takes some 10^{10} years to generate and radiate this amount of energy, the estimate of the total energy radiated by the most powerful radio galaxies is based on the assumption that the process lasts for only about a million years. Some extremely violent explosive mechanism would therefore seem to be indicated.

This conclusion has a bearing on the problem of the origin of the most energetic cosmic ray particles. For it seems likely that in a titanic explosion of a galaxy (as distinct from the supernova explosion of a star) some of the particles involved might well fly off into intergalactic space with energies up to 10^{19} electron-volts, or even higher.

The exploding-galaxy explanation of radio galaxies gained support from the discovery, beginning with the case of Cygnus A in 1953, that many radio galaxies can be resolved into two separate regions of radio emission. In the case of Cygnus A it was found that these regions are located on opposite sides of the visible galaxy at distances of about 100 000 light-years. This discovery suggested that the twin radio regions represent two jets of high-energy particles ejected by the galaxy and moving in opposite directions so as to conserve the linear momentum of the system as a whole. They would carry the galaxy's magnetic field with them. Synchrotron radiation would occur all along the jets, but would be most concentrated at the ends where the lines of force of the magnetic field would be most compressed.

However, it was not until 1961 that convincing optical confirmation of the possibility of explosions on the galactic scale was obtained. In that year a small neighbour, M 82, of the giant spiral galaxy M 81 (outside the local group and some ten million light-years distant from us) was found to be a discrete radio source. Old plates of M 82 showed extensive dust lanes and a faint filamentary structure extending above and below the main disc of the galaxy. Stimulated by the discovery that this object was a radio source, Sandage photographed it in 1962 with the 200-inch telescope, and took the important step of using a special filter that admitted only red light of the hydrogen α-line in order to detect any prominent hydrogen structures that otherwise might have escaped detection. The result was dramatic. What had previously been detected as inconspicuous filamentary wisps were now seen to be vast spectacular tongues of glowing hydrogen extending more than 14 000 light-years above and below the main disc. Moreover, careful spectroscopic investigation revealed evidence of motion away from the hub of the galaxy along the axis of rotation. Since the velocities were proportional to the distances from the hub (at the ends of the filaments they were about 600 miles a second), it followed that all the matter in the filaments must have been back in the nucleus of M 82 at a given time in the past, some one and a half million years before the stage we now see. This is convincing evidence that the filaments are due to a single vast explosion that occurred in the nucleus of the galaxy. From the strength of the main hydrogen emission line, the amount of material involved was found to be about 5 000 times the mass of the sun, or about one part in 2 000 of the mass of M 82 itself.

Never before had evidence been found that explosions can occur in nature on so vast a scale. One of the puzzling features was the fact that all this hydrogen gas is ionized. Interstellar hydrogen in our galaxy is ionized only when in the neighbourhood of hot blue stars, but M 82 contains no blue stars in its filaments. The explanation was found by observing that the light from the filaments is highly polarized, with the electric vector of the light parallel to the plane of the galaxy. From this it appears that M 82 possesses a magnetic field aligned along the axis of rotation, and this presumably generates by the synchroton mechanism enough ultra-violet radiation (as well as the observed radio flux) to ionize the hydrogen gas in the filaments, provided the explosion produced enough electrons at the required energy (10^{16} electronvolts).

Cosmic ray particles must be very abundant in M 82. An observer on a hypothetical planet orbiting a typical star in this galaxy would be able to detect a local cosmic ray flux of electrons that is about a thousand times as intense as in the neighbourhood of our earth, and cosmic ray protons must be present in great profusion.

The total energy needed to set the exploding matter in M 82 in motion is about 2×10^{48} joules, although to produce this kinetic energy an initial energy input of perhaps 10^{50} or 10^{51} joules may have been required. The filaments have counterparts in other radio galaxies, such as the jet in M 87, and M 82 may be regarded in this respect as a more or less typical radio galaxy. But what of the most powerful radio galaxies such as Cygnus A, for which, as we have seen, an energy input of 10^{54} to 10^{55} joules seems to be required? Even if the conversion of mass to radio energy were one hundred per cent efficient, at least 10^9 to 10^{10} times the mass of the sun would be required, corresponding to the entire mass of a medium-sized galaxy. Allowing for conversion efficiencies of the order of one per cent, a mass of the order of 10^{12} times that of the sun, and therefore comparable with the very largest galaxies, would seem to be required!

Quasars and Quasi-Stellar Galaxies

During the 1950s it came to be generally accepted that most of the discrete radio sources would in due course be identified with galaxies. Improvements in technique, in particular the development of the radio interferometer at Jodrell Bank, led to resolving powers of less than one second of arc. It then appeared that some strong radio sources were unusually small. Three that were thought to be remote galaxies were photographed by Sandage and Matthews with the 200-inch telescope, but when pictures of the search areas concerned were studied no trace of any galaxies could be found. Instead, there appeared to be a faint star close to the radio positions. When the colours and spectra of these 'stars' were examined it was found that each was generating an unusually large amount of blue, violet and ultra-violet radiation. Their optical spectra showed no known spectral lines.

Shortly afterwards, in the autumn of 1962, Australian radio astronomers led by C. Hazard succeeded in pinpointing with great accuracy the position of the radio source 3C 273 as a result of its occultation by the Moon. It was found that the source consisted of two small components, one of which coincided with a bright blue star and the other with a fainter wisp or jet of bluish nebulosity (see Plate XXVIII). As the star was relatively bright (12th magnitude), Maarten Schmidt of Mount Wilson and Palomar obtained its spectrum and found six spectral lines which at first he could not identify. Then one evening it occurred to him to see what would happen if a red-shift or blue-shift correction were applied to them. To his surprise he found that, if the spectrum were corrected for a red-shift of 0·158, all six lines could be identified. Similarly, when the spectrum of 3C 48, one of the 'stars' investigated by Matthews and Sandage, was re-examined, the lines in its spectrum could also be identified if allowance was made for a red-shift of 0·361.

These unexpected results implied either that the objects in question were extremely massive stars in our galaxy with gravitational fields intense enough to cause these large red-shifts, or else these shifts might be due to the general expansion of the universe, in which case the objects must be more than a thousand million light-years away. In either case it seemed that a new class of astronomical object had been found, and the term 'quasi-stellar radio source', later abbreviated to *quasar*, was introduced for them.

The arguments against quasars being local objects are strong. For, if they are stars of mass of the order of magnitude of that of the sun, then to produce the required red-shifts they must be of very small radius, of the order of 10 kilometres, and so possibly neutron stars. But, in this case, it is extremely difficult to explain their spectra. These contain emission lines which could only arise in a gaseous shell a few hundred metres thick, for otherwise they would be broader than observed because of differences in the gravitational red-shift at different distances from the centre. The occurrence of forbidden lines in the spectra implies low particle densities in the outer regions and hence the objects must be relatively poor emitters, in which case their actual brightness could only be explained if they were a few tens of kilometres away from us! A distance of the order of 6 000 000 light-years could only be attained in this way if the mass were of the order of 2×10^{13} that of the sun, making these objects easily the most massive systems in the universe! In any case, the later discovery that 3C 273 lies behind the Virgo cluster, since it shows absorption at the 21 centimetre line, presumably from hydrogen gas in that cluster, is a positive indication that quasars are at cosmological distances.

If we interpret the red-shifts of quasars as due to the general expansion of the universe, as most astronomers are now convinced we must, we are still faced with serious problems. (The argument that quasars may have been shot out from a neighbouring galaxy is now generally rejected. The high velocities and the fact that they all appear to be recessional make this hypothesis most implausible.) For their apparent magnitudes imply that they emit up to one hundred times as much light as the brightest

ATOMS AND THE UNIVERSE

galaxies known, and are therefore by far the brightest optical objects in the universe, although they do not emit so much radio radiation as the most powerful radio galaxies. Nevertheless, some of them must be very much smaller in volume than galaxies. For not only do they present a star-like appearance, but it has been found that they tend to vary in brightness, including one case in which there was a two-fold change of brightness in twenty-four hours. From this it follows that the diameter of the object cannot be more than a light-day, and in other cases of the order of a light-month, which is only about a hundred times the diameter of the solar system.

The most striking characteristic of the optical spectrum of quasars is the strong emission in the ultra-violet. This property has been used to identify quasars optically when the radio position is sufficiently ambiguous to include a large number of star-like objects. Two plates are taken of the region involved, one with a blue filter and the other with an ultra-violet filter, the exposure times being adjusted so that the images of normal stars will appear equally bright on both plates. It is then possible to identify quasars by their relatively stronger images on the ultra-violet plate.

Using this technique Sandage found, in 1964, a number of objects which resembled quasars optically but were not detectable emitters at radio wavelengths. These blue compact objects are known as *quasi-stellar galaxies*. They are believed to be somewhat more numerous than quasars, which are much less numerous than galaxies. They may represent a later stage of quasar development in which the radio emission is no longer detectable.

One of the most intriguing features in the spectra of quasars is that in some there are absorption lines that are red-shifted by different amounts from each other and from the emission lines. Presumably, absorption occurs in rapidly moving shells of gas that have been ejected by the quasar.

Many hundreds of quasars have been discovered and it is now realized that, although some of the earliest found were selected on the basis of their very small radio angular size, these were exceptional. Many are double radio sources on a scale comparable with the most extensive radio galaxies. In some cases there is a third radio component that coincides with the visual object, and variations in its radiation indicate a highly compressed small-scale source. It seems that the enhanced optical emission probably persists for about a million years, so that the total energy involved is about 10^{53} joules.

The Evolution of Galaxies

Other objects which seem to radiate large quantities of energy in a comparatively short time are the spirals known as *Seyfert galaxies*, after Carl Seyfert who first studied them in 1943 (see Plate XXIX). They are thought to comprise about one per cent of all galaxies. They differ from normal spirals in having very small but intensely bright nuclei containing some 10^9 or 10^{10} stars within a radius of about 1 500 light-years. The spectra of these nuclei reveal broad emission lines, indicating that the atoms present are in a high state of excitation. It has also been found, in at least two cases, that the light emitted varies markedly in a period of months, and several are strong emitters in the infra-red. Indeed, many of the characteristic features of the nuclei of Seyfert galaxies are similar to those of quasars.

Some powerful radio sources are associated with the so-called N-galaxies, which contain a compact nucleus like the Seyfert galaxies but are much more luminous. In these sources energy production is still occurring. Some comprise a double structure with an age of more than a million years, confirming that activity within the nucleus can continue for this length of time.

It is now thought probable that there is some intimate connection between strong radio galaxies, quasars, the nuclei of Seyfert galaxies and the whole range of violent activity found in the nuclei of spiral galaxies generally, as well as jet phenomena such as that observed in the massive elliptical M 87. In particular, as Sir Martin Ryle has stressed, observations reveal the existence of a wide range of objects all of which seem to require a minimum total energy of about 10^{53} joules, although the time scale for the release of this amount of energy varies considerably from one class of object to another. In the case of Seyfert galaxies, M 87 and other relatively weak radio sources, energy production may continue for about 100 million years. For some quasi-stellar sources the duration may be more than a million years, whereas for the radio galaxies most of the energy is probably produced in only 100 000 years. The rate at which energy becomes available seems to be the decisive factor in determining the type of object produced.

The development of a galaxy into a quasi-stellar source, radio galaxy, N-galaxy, Seyfert or normal galaxy may well depend on the rate at which non-thermonuclear energy is made available. If a typical galaxy is formed from a primeval gas cloud, its mass, angular momentum, and the scale of irregularities initially present in it may be the key factors controlling its subsequent evolution. The proportion of galaxies a that are likely to become, say, Seyferts can be determined, assuming that it is independent of epoch, from the formula

$$\rho = a\rho_0 T/T_0,$$

where ρ is the local density of Seyferts, ρ_0 is the local space density of all galaxies, T is the period for which a galaxy is recognizable as a Seyfert, and T_0 is the age of the universe. It is found that, whereas a large proportion of galaxies can become Seyferts, only a small proportion will become powerful radio sources or quasi-stellar objects, if the relative number-densities of the different kinds of galaxy remain the same at all epochs. However, this condition does not appear to be satisfied, since at the epoch when it seems galaxies may have formed (corresponding to a red-shift of 2·5, i.e. rather more than 8 000 million years ago), observations indicate that the relative space density of the most powerful sources must have been a thousand times the present value. The observations in fact seem to show that comparable numbers of galaxies develop into the different kinds of source mentioned above, but those galaxies that exhibit a rapid release of energy are most likely to manifest this effect at an early stage in the evolution of the universe.

The Primeval Fireball

In 1965, evidence of a still earlier stage in the evolution of the universe than that already revealed by radio galaxies and quasars was unexpectedly discovered by A. A. Penzias and R. W. Wilson of the Bell Telephone Laboratories. In testing a sensitive microwave radiometer they found that some radiation appeared to be leaking into the antenna of their apparatus. They soon discovered that the source of this radiation was more or less isotropic and that at the wavelength at which they were working, 7·4 centimetres, the radiation appeared to be equivalent to that of a black-body at about 3·5 K. Later this value was corrected to about 2·7 K.

Meanwhile, R. H. Dicke and his colleagues at Princeton University were arguing on purely theoretical grounds that a universal microwave radiation with similar properties should exist. Their argument was based on the idea that in an evolutionary universe that originally 'exploded' from a highly compact source some eight to ten thousand million years ago the initial temperature must have been extremely high. As the universe expanded out of this inferno, the matter cooled to form galaxies and stars, but the primeval radiation which had started out as enormously energetic γ-rays was 'cooled' by the expansion and its wavelength increased so that today it now appears mostly in the radio and microwave bands of the spectrum. The idea of this primeval 'fireball' is not one of radiation coming to us merely from some remote source. On the contrary, we are immersed in it, and any observer anywhere in the universe at the same local time as we are at now would observe this radiation coming equally from all directions at the same temperature and with the same other properties. Strangely enough, Dicke and his colleagues were quite unaware that, as recently as 1949, George Gamow and others had predicted, on the basis of a hypothetical explosive origin of the universe, the existence of cosmic background radiation with a present temperature of about 5 K and a black-body spectrum.

The general idea of the evolution of the universe implied by these theoretical investigations may be briefly summarized. At a temperature of 10^{10} K, the densities of matter and radiation are about equal, being about 10^6 to 10^7 grammes per cubic centimetre. Rapid evolution follows with radiation densities at first predominating. But after 10^3 to 10^4 years the density of matter starts to exceed that of radiation, and this continues until the present epoch when it is overwhelmingly predominant. At an early stage deuterium would be formed, but only when the temperature is low enough for it to be stable against photo-disintegration. If the density is sufficiently high at this stage, large amounts of helium will be generated, but this reaction will cease when the temperature falls. Hence, if we can determine the present helium content of the universe, an upper limit to the matter density of the universe at the time of helium formation can be calculated. Unfortunately, there is considerable controversy concerning the present helium content of the universe, and the usefulness of this line of investigation for helping to decide on the most appropriate cosmological model to describe the evolution of the universe cannot yet be fully exploited.

Nevertheless, one very important consequence is already clear. If this primeval fireball interpretation of the microwave background radiation is accepted, then the steady-state model of the universe must be finally rejected, since this universe was, by definition, never in a significantly denser state than it is at present and therefore could not have generated the observed radiation.

The best confirmed feature of the microwave radiation is its isotropy: independence of direction has been established to within 0·5 per cent in a circle near the equator. This is sufficiently precise to exclude the possibility of any local origin for the radiation, since a source restricted to the solar system, the galaxy, or even the local group would not appear isotropic to an observer located, as we are, far away from the centres of these systems. The test of isotropy provides a necessary, but not a sufficient, condition for the primeval fireball hypothesis.

The small departure from absolute isotropy, with its daily variation as the earth rotates, can be used to set an upper limit on the component velocity of the solar system (in the equatorial plane) relative to the general background of the universe. It has been found in this way that this component of our motion cannot exceed 300 kilometres a second, a result which may be compared with the velocity of the solar system due to the rotation of the galaxy, which when projected onto the same plane is about 200 kilometres a second.

The Origin of the Elements

Throughout the universe there is on the whole remarkable uniformity in the relative abundances of the elements. Besides predicting the existence of cosmic background radiation, in their pioneer paper of 1949 Gamow and his colleagues sought to explain this uniformity on the assumption that the universe began as a hot nuclear gas at a temperature of thousands of millions of degrees so that all matter was in the form of the simplest particles – protons, neutrons and electrons. Owing to expansion, this gas became both less dense and much cooler and neutrons became attached to protons to form complex nuclei, the prototypes of the atomic nuclei of today. In order to account for the observed relative abundances of the different elements, the conditions of temperature and pressure at this stage would have required delicate adjustment. The principal difficulty of the theory, however, arises as a direct consequence of the fact that it entails a step-by-step process of successive neutron capture to build up the more complex nuclei from the simplest. Owing to some peculiar interplay of nuclear forces, neither a single proton nor a single neutron can be attached to the helium nucleus, of mass 4, to obtain the next nuclide, of mass 5. There is, in nature, no nuclide of mass 5, and although it has been produced artificially in the laboratory it immediately disintegrates. Hence, to build up a nuclide of mass 6, two particles must be captured simultaneously by a helium nucleus. Unfortunately, under the assumed physical conditions, the probability of this happening is negligibly small. It is possible that this difficulty may eventually be overcome, without special *ad hoc* hypotheses, by more detailed study of the self-heating of the nuclear gas, but the calculations are likely to be extremely complicated.

A radically different theory was later developed with great ingenuity by W. A. Fowler, F. Hoyle and others on the hypothesis that all heavier elements are generated in stars. In particular, Hoyle has shown how, in the interior of a star with a hot core of helium at a temperature of a hundred million degrees, nuclei of mass 8, which are as ephemeral as those of mass 5, can be produced as rapidly as they decay. It is therefore possible from time to time for such a nucleus to fuse with a helium-4 nucleus to produce carbon-12. Hoyle found that a carbon-12 nucleus generated in this way would have certain peculiar properties. Subsequent laboratory experiments confirmed that carbon can in fact exist in the required state. Other elements with atomic weights which are multiples of four, up to iron-56, can be formed by the successive addition of a-particles, at temperatures of the order of a hundred million to a thousand million degrees. But, in order to generate nuclei of atomic weights which are not multiples of four, it is essential to have hydrogen nuclei available in their vicinity. It is still an open problem whether at the very high temperatures concerned sufficient hydrogen can be kept from turning into helium. Be that as it may, the stellar production of some nuclides can only occur under the special conditions associated with supernovae. Indeed, supernovae explosions play a central role in this theory, for if heavy elements are produced only in stellar factories then some ejection mechanism must be invoked to explain their presence in interstellar space.

That the stars synthesize at least some heavy elements has been confirmed observationally in various ways. The most remarkable evidence concerns the unstable element technetium. Its longest-lived isotope has a half-life of just over 200 000 years – far less than the age of the stars in which its spectral lines have been observed. To exist in a significant quantity this element must therefore have been generated in the star long after the star itself was formed.

It is possible, however, that there may be an entirely different explanation for the origin of most heavy elements in some primeval explosion at the centre of our galaxy, as has been suggested by V. A. Ambartsumian.

The Origin of Stars

Whether the galaxies and clusters of galaxies have condensed out of a universal primeval gas at an early stage in cosmical evolution, is still an open question. There is somewhat more evidence bearing on the origin of stars than on that of galaxies, although it seems probable that the two problems are closely linked. At the present time, we believe, star formation is still going on in irregular galaxies and the arms of spirals, which are Population I systems. These systems contain large quantities of dust and gas from which stars may be generated by condensation, probably in groups.

It has been suggested that a spiral galaxy may be an irregular galaxy rotating within an elliptical galaxy and thereby swirled into the spiral form. As already mentioned, when a typical spiral, such as the Andromeda nebula, is photographed in the infra-red, the spiral shape is no longer seen and the galaxy appears to be a vast system of spheroidal shape. This strongly suggests that the spiral arms, which appear so prominent in the usual photographs because they contain the most luminous stars, are minor attachments to the main body of the system in which they occur. Moreover, they are probably comparatively short-lived. Study of the spectra of light from different regions in the arms shows that they revolve about the central core with speeds depending on distances from the centre. The outer regions revolve more slowly than the inner regions, just as in the solar system the outer planets revolve more slowly around the sun than do the inner planets. Hence, if spiral arms persisted for several revolutions, they would tend to become twined up. In fact, the arms usually show only one or two complete turns, indicating that, in all probability, they are young, transitory features of the galaxies in which they occur.

This conclusion accords with the fact that, being Population I systems, spiral arms are dominated by highly luminous energy-spendthrift stars. Indeed, some super-giant stars have a total life-span of the order of only a million years and must, therefore, have been formed very recently. It is highly probable that stars like these are still being formed in our part of the Milky Way. For the emergence of new stars, we must look to the thickest dusty regions, for example the great cloud complex of which the Orion nebula is the most conspicuous feature (see Plate XX). The creation of new stars may actually be in progress in very dark parts of this region.

Recently, the discovery of the highly energetic nature of the nuclei of galaxies and of the explosive phenomena associated with them, has led to a revival of Jeans's suggestion, made in 1928, that the centres of galaxies may be 'singular points at which matter is poured into our universe from some other, and entirely extraneous, spatial dimension so that, to a denizen of our universe, they appear to be points at which matter is being continually created'. If this is so, then the existence of such 'white holes' (the reverse of the 'black holes' mentioned on p. 230) in the universe might be an alternative means of accounting for the original formation of stars, and possibly galaxies. There is evidence that much matter, including stellar groups, may be ejected violently from galactic nuclei. On the other hand, attempts to detect intergalactic hydrogen, e.g. by absorption of the ultra-violet lines in the spectra of very remote objects in the universe, have so far led to surprisingly low densities being assigned to this material. If this diffuse matter is so sparsely distributed in space, it is difficult to understand how its condensation into galaxies and stars has been so efficiently effected. The whole problem is, of course, wide open, and all that can be said at the present time is that the traditional belief that stars and stellar systems originally condensed from masses of nebulous material spread out over vast regions of space may be a fallacy. Instead, both the universe itself and its major constituents (stars and galaxies) may all have had explosive point-like source origins.

Stellar Evolution

The more massive a star, the more prodigal it is in burning up its available hydrogen. Thus, whereas the sun has consumed about 8 per cent of its hydrogen in the past 5 000 million years, the life-span of Sirius cannot be more than about 1 500 million years in all. The sun is therefore a much older Population I star than the more massive Sirius, which in its turn is probably older than an extremely luminous star such as Rigel.

The five divisions of stars which have been suggested in place of Baade's two discrete populations are believed to correspond, more or less, to the sequence of stellar ages. The comparatively rare blue super-giant stars and the stars in galactic clusters are all young and belong to Extreme Population I. Most other stars in our region of the galaxy belong either to Intermediate Population I or to the so-called Disc Population. In the former the spectral lines due to metals are stronger than they are in the latter. Presumably, this means that the abundance of heavier elements in the Milky Way was somewhat greater when the stars of Intermediate Population I were formed than when those of the older Disc Population were generated. The abundance of heavier elements is least in Population II stars, particularly in those of Extreme Population II found in globular clusters. These are believed to be the oldest stars and were formed when the chemical composition of the galaxy was most primitive.

Although, according to current theory, the luminosity of a star when born is governed primarily by its mass, it also depends on the chemical composition. Thus a star which is deficient in the heavier elements will be only about half as bright as a star of the same mass that has a chemical composition similar to that of the sun. For stars of the same composition the initial luminosity varies as the fourth power of the mass. In theory, if the surface temperatures (or colours) of a cluster of zero-age stars of similar composition are plotted against their absolute magnitudes, they are presumed to lie along a straight line (strictly speaking, a band of narrow width) running from faint-and-red to bright-and-blue. In practice, if we plot on a colour-magnitude diagram of this type points determined by observations of the stars in an actual cluster, we find that they usually lie along a line which merges with the theoretical zero-age line only towards the faint-and-red end. As we pass to brighter stars, sooner or later the line tends to bend away from the blue (high surface temperature) back towards the red (lower surface temperature). This is believed to be an evolutionary phenomenon, the brighter stars converting their available hydrogen more rapidly into helium and hence tending to diverge sooner from their state at zero age. Consequently, the position of the turn-off point, or bend, in the colour-magnitude diagram of a stellar cluster is taken to be a good indication of its age.

It is thought that normally in a star there is little mixing of material between the hot central core and the cooler layers above. When the hydrogen supplies in the former begin to peter out, the star must change rapidly until its interior is sufficiently hot for a new type of nuclear reaction to occur – the conversion of helium into carbon. We are not yet clear about the precise order of events when, in its turn, this process begins to fail, but it is probable that the star eventually explodes violently in a Type I supernova outburst, distributing much of its contents into interstellar space. Presumably, the remnant star then progresses slowly through the white-dwarf stage until a final black-dwarf state of total darkness and extinction is reached. The fact that white dwarfs appear to be very numerous lends support to this hypothesis, but there are reasons for believing that the white-dwarf state is not possible for masses exceeding that of the sun by more than about 50 per cent. Instead, for more massive stars it has been conjectured that eventually their central regions are converted into iron and neighbouring elements in the Periodic Table. When this happens no more energy can be released by nuclear reactions in these regions but their temperature continues to rise. A rapid collapse then releases much gravitational energy and the outer regions, which still contain potential nuclear fuel, rise to a very high temperature and react explosively to produce a Type II supernova. Such an explosion probably produces heavy elements, which are ejected into space, and also cosmic rays. The remnant of the star may then become a pulsar, like that at the heart of the Crab nebula. Ordinary novae are now thought to be partners in close binaries, the nova outburst accompanying a rapid exchange of mass between the two stars.

The Laws of Nature

Surveying the universe from the solar system out to the farthest depths of space, scientists have boldly applied the fundamental laws of physics to cover immense distances of space and correspondingly enormous stretches of time. The extrapolations from direct observation and laboratory experience which are involved have been described by a distinguished American physicist and philosopher of science, P. W. Bridgman, as 'hair-raising'. With what degree of confidence, then, can we accept the pronouncements of modern astronomers and cosmologists on the structure of the universe, and in what way, if any, are their views to be regarded as better founded than those which were held in previous centuries by men of equal intelligence?

In a famous lecture on the origin of the solar system, delivered over eighty years ago when knowledge of the structure of matter and of stars was far less extensive than today, the great German physicist and physiologist Helmholtz justified the study of the subject despite its predominantly speculative nature. He argued that, far from avoiding problems of cosmogony, scientists were not merely entitled to consider these problems but should regard it as their duty to study them. They should investigate whether 'on the supposition of an everlasting uniformity of natural laws, our conclusions from present circumstances as to the past . . . imperatively lead to an impossible state of things; that is, to the necessity of an infraction of natural laws, of a beginning which could not have been due to processes known to us'. Considered as a question of science, this, he maintained, was no idle speculation, for it concerned the extent to which existing laws are valid.

The principle of extrapolation of the laws of nature which have been empirically tested in the laboratory and by observations on the solar system is often called the 'principle of the uniformity of nature'. The crowning achievement of Isaac Newton was to provide compelling arguments for extending laws governing physical phenomena here on earth to the heavens. For two thousand years the contrary opinion of Aristotle had prevailed. His separation of the terrestrial and celestial realms was, however, not the purely arbitrary imaginative hypothesis that it is nowadays so often made to appear. It was partly based on sober empiricism. 'It is absurd', he wrote, 'to make the universe to be in change because of small and trifling changes on earth, when the bulk and size of the earth are surely as nothing in comparison with the whole universe.'

The extrapolation of physical laws from the terrestrial scale to the cosmical is both a hypothesis and a method of investigation which has continually to justify itself by results. So far from being mere 'common sense', it has on occasion led great men of science to accept the most fantastic conclusions. An outstanding example was Sir William Herschel's belief in the habitability of the sun.

The Copernican conception of the earth as a planet provided a powerful argument for believing that similar physical processes to those occurring on earth prevailed on the other planets, and it was natural to conclude that since the earth is a habitable globe the other planets must be too. Nowadays we discriminate more carefully between the uniformity of physical laws and of physical conditions. The belief in the uniformity of the universe and the plurality of worlds led Herschel, in a memoir *On the Nature and Construction of the Sun and Fixed Stars* published by the Royal Society in 1795, to argue that the sun is an overgrown planet with an abnormally luminous atmosphere. Below this atmosphere the main body of the sun was, in his opinion, opaque and of great solidity. This idea was not pure speculation. It was based on the observation that sun-spots appear dark against the general bright background of the solar surface. We now believe that this is because the spots are regions of somewhat lower temperature, but Herschel suggested that they were regions where we look right through to the solid surface beneath the sun's atmosphere. Moreover, he thought that this surface was diversified with mountains and valleys. 'I think myself authorized,' he wrote, 'upon astronomical principles to propose the sun as an inhabitable world.' He based this conclusion on the essential similarity of the sun and planets! 'Its similarity to the other globes of the solar system with regard to its solidity, its atmosphere and its diversified surface, the rotation upon its axis, and the fall of heavy bodies, leads us to suppose that it is most probably also inhabited, like the rest of the planets, by beings whose organs are adapted to the peculiar circumstances of that vast globe.'

To meet the common-sense objection that, in view of the heat which we receive on earth from the sun, the inhabitants of the sun's solid surface could not be protected by a mere blanket of cloud from being roasted, he again appealed to the uniformity of natural laws. He pointed out that high up in our own atmosphere, where there is less to impede the sun's rays than on the earth's surface, it is actually colder. Writing only eleven years after the first successful aeronautical ascent in 1783 and in the same year in which the first military balloon was used, by the French for reconnaisance before the battle of Fleurus, Herschel not only refers to the temperature on mountain tops but also to the fact that 'our aeronauts all confirm the coldness of the upper regions of our atmosphere'. Thus, arguing from correct observations and plausible hypotheses in a manner which he was convinced was strictly scientific, the greatest astronomer of his age drew conclusions concerning the nature of the sun which were no less fantastic than Aristotle's idea that it was a 'perfect body' in which change and decay could never occur.

Encouraged by Helmholtz's exhortation, but also warned by the example of Herschel's fallacious application of the principle of uniformity, how are we to assess the far-reaching conclusions concerning the physical universe drawn by contemporary astronomers and cosmologists? Of the hundred thousand million stars which form our galaxy, only the sun is sufficiently near to us for direct observation of surface details. Our ideas concerning the surfaces of all other stars, as well as of their internal structures, are inferences dependent on our conception of physical laws. When we come to consider the universe as a whole and in particular its origin and age, fundamental questions concerning the nature and scope of scientific method and of the laws of nature force themselves upon us.

Newton was fully aware of this aspect of the cosmological problem and of the hypothetical nature of the principle of uniformity. In one of the Queries inserted at the end of his *Opticks* he was careful to point out, 'It may also be allow'd that God is able to create Particles of Matter of several sizes, and in several Proportions to Space, and perhaps of different Densities and Forces, and thereby to vary the Laws of Nature, and make Worlds of several sorts in several parts of the Universe. At least, I see nothing of contradiction in all this.'

Nevertheless, although astronomers can give no final answer to the vast problems of cosmology and cosmogony, in particular the problems of the origin and age of the universe, they now stand at a much better vantage point than did their predecessors. Of course, it is still just as true now, as when Newton wrote, that the principle of uniformity of natural laws is a hypothesis and it is just as true now, as when Helmholtz wrote, that no theory concerning the origin of the world can be verified by direct observation, yet it seems remarkable that so many independent lines of enquiry into the history of the earth, the stars and the galaxies yield ages of the same order of magnitude, all being in the region of five to ten thousand million years. But whether this is to be taken as an indication of the time that has elapsed since the whole physical world was created or only since it took the form which we now study is a question which takes us beyond the limits of scientific enquiry. 'Our conclusions from present circumstances as to the past' lead us to regard this as the longest stretch of past time over which we can extend the laws of nature as we know them. Beyond that we cannot penetrate.

Appendix

Theories and the Role of Mathematics

It is generally known that mathematics and physics are intimately bound up. Every physicist has to have training in mathematics, and a practising physicist may use mathematics as often as he uses a measuring instrument. The non-mathematical reader cannot be expected to follow the mathematical arguments with which the physicist works, but he can gain some understanding of the role of mathematics in physics.

First, he should understand that mathematics is not a method by which the scientist 'draws truth out of the sky'. It is, in the first instance, a shorthand by means of which the results of experiments can often be expressed concisely. For example, one of the most important of physical laws is Newton's second law of motion

$$F = ma,$$

where F stands for the force acting on a body, m for its mass and a for its acceleration, or rate of increase of velocity.

The formula given above makes use of algebraic symbolism. Thus F, m and a really stand for 'hidden' numbers, like 'x' in elementary algebraic expressions. The significance of a given m, for example, is that it represents that number of the particular units which have been chosen as convenient for the measurement of mass. The units must be chosen to be consistent with each other; then by the use of this formula we can find what would be the acceleration of a body of known mass under the action of a given force.

But mathematics has a wider application than we have so far

illustrated. For there can be formulae of all kinds in physics, relating more and more physical observations. By making use of the well-known techniques of algebra we can sometimes establish links between several formulae, and perhaps arrive at conclusions quite distant from those which could be established in any given experiment. The Thomson experiment which was discussed in the Introduction contains quite a good example of this. In fact, it consisted of two experiments; in one, the effects of magnetic and electric forces were 'balanced', as already described. In the second, the actual deflection due to a lateral magnetic force was measured.

Considering the first experiment only at present, we have to make use of the known physical fact that the force upon a particle carrying an electric charge e moving with velocity v at right angles to the direction of the field of a magnetic force H is equal to Hev.

Similarly for the effect of electric force, we know that the force on a charge e in a field of electric force E is Ee. If the two forces are balanced, we therefore have

$$Ee = Hev,$$
and so
$$v = E/H.$$

The second experiment enabled Thomson to find the value of e/m. It was necessary to make use of another known physical fact: that a force F acting at right angles to the direction of motion of a particle of mass m moving with velocity v deflects it along a circular path of radius r, such that $F = mv^2/r$, whence

$$r = \frac{mv^2}{F}.$$

In this case, we know that $F = Hev$, so that $r = \dfrac{mv}{He}$.

By combining this with the previous formula, again using ordinary algebraic operations, we find

$$\frac{e}{m} = \frac{E}{H^2 r}.$$

The value of r could be worked out, using further simple mathematical arguments, from the value of the deflection observed and the length of the path of the beam between the parallel plates, and hence a value obtained for e/m. Incidentally, it will be noted that v 'drops out' of the argument. It is not in itself a fundamental property of the electron but depends upon the voltage applied between the terminals of the discharge tube.

The degree of complication in modern physics requires the use of mathematics which is much more difficult than the elementary algebra used above. In particular, it involves constant use of a powerful mathematical tool known as 'calculus'. Its scope can be illustrated by an example. It is easy to work out the velocity at any time of a falling body. For the acceleration, or rate of increase of velocity, of a falling body on the earth is a *constant* quantity, 981 centimetres per second per second, denoted by g. Then after time of fall t the velocity v is given by

$$v = gt.$$

But how can one work out the *distance* fallen after a given time, remembering that the velocity (that is, the rate of traversing distance) is not constant but increasing continuously? The answer can be found graphically. The variation of the velocity with time is correctly represented by a graph of the form shown in Fig. 29 (a) in which v is plotted against t, that is, the velocity

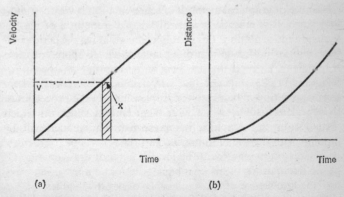

(a)   (b)

Fig. 29. (a) *Velocity of a falling body;* (b) *Distance travelled by a falling body*

increases uniformly (or, as we would say, *linearly*). During a small interval of time x, the distance fallen is nearly equal to xv, where v is the velocity at, say, the beginning of the interval. It can thus be represented by the shaded area. Clearly, therefore, the distance fallen at any time can be represented by the total area under the line up to that time, when appropriate consideration is given to the matter of units. If we measure this area we find, for example, that the distance fallen is related to the time by a graph of the form shown in Fig. 29 (b).

The function of the calculus is to extend the scope of algebraic symbolism and operations to cover such problems as the derivation of the form of one of these graphs from the other, or vice versa, without the necessity for, say, the actual measurement of areas. By the use of such methods we should be able, for example, to work out the motions of electrons in tubes containing complicated assemblies of plates which need not be parallel, under circumstances in which the magnetic or electric forces were not fixed but varying, as in alternating current. This is indeed one of the main problems in electronics. It was by such an approach that methods were devised during the war for generating powerful beams of radio waves of wavelength sufficiently short for use in radar; that is, fairly short compared with the size of the object – the aircraft – being looked for.

The use of graphical methods of representation is very common in science. It is essentially a corollary of the method of mathematical representation. For example, if certain new experimental observations are made it is often much easier to represent them by drawing a graph than to find an appropriate mathematical formula. It is significant that at scientific conferences, where the presentation of new results occupies much of the proceedings, the speakers usually bring with them lantern slides on which their results are graphically represented. The reader will be familiar with the use of graphs in, say, the reports of companies, or in election manifestos. If he has the habit of drawing a graph of the fluctuations of his own bank account he may have a very realistic knowledge of the significance of negative values.

But mathematics is even more important in physics than this. Not only can the mathematician deal with problems, such as those just mentioned, concerned with the interpretation of experiments or the forecasting of behaviour in particular pieces of apparatus; he can also treat natural objects as pieces of apparatus. For example, knowing as he does that a hydrogen atom consists of a negatively charged electron and a positively charged nucleus (or proton), he can apply the formulae already discovered in ordinary laboratory experiments, relating the motions of charges with the electrical and magnetic conditions, to try to predict the behaviour of hydrogen atoms. The results of such enquiries might agree with the experimentally observed facts about the behaviour of hydrogen atoms. (Actually they do not, as we have seen.) If they did, he would decide that his idea of the structure of the atom was more or less complete, and that its constituent parts obeyed the same rules as do the larger objects studied in ordinary laboratory experiments.

In physics today there is a continuous interplay between mathematics and experiment of the kind implied here. Indeed, mathematics in physics has become so important, and so complex, that there now exist mathematical or theoretical physicists who are expert in mathematical techniques rather than in experimental techniques. Sometimes discoveries are, in a sense, first made by the theoretical physicist and later confirmed by the experimenter. Probably the most famous example of this was the discovery by J. Clerk Maxwell that the general formulae connecting electric and magnetic phenomena, when combined, appeared to indicate that electromagnetic waves must exist, before radio waves had been experimentally discovered. On other occasions the experimenter first makes his observation, and it is later explained by the theoretician. Of course, the ordinary experimenter also tries to be a theoretical physicist as far as his mathematics and his opportunities allow, and he can usually cope with the simpler situations. For example, he is usually adept at spotting whether results expressed graphically imply some simple mathematical relationship between the quantities under consideration.

Of course, the role of mathematics is not confined only to the development of pure physics. As already illustrated, mathematics may be used to predict behaviour when no uncertainty of *principle* exists. It is easier and cheaper to perform the mathematical calculation than to build a model and subject it to an experimental test. For example, provided the basic strength of the steel used is known, and the strengths of the joints, it will be possible to calculate the strength of a bridge made up of a lattice of steel girders. An experimental test would be expensive. Equally, if a bridge is required for a particular purpose, that is, to bear a certain load, or to have a given strength, then the choice of design – the structure of the lattice and the sizes of the various members – may be selected with the use of mathematics. This would be described as a piece of *applied science*, or *engineering science*.

As we have seen, the use of the computer in mathematics is to increase the *speed* of calculation. It does not alter the mathematics. In pure science, calculations can be performed which could not be performed in any other way because they would take too long. The reasons for the applications of computers in applied science are often the same. We have all noted the dramatic use of computers in controlling the moon landings. It might perhaps have been possible to calculate by conventional means what changes in direction were necessary in mid-flight to bring the space-craft into a suitable approach path. But the closer approach, and the last moments in particular, could not have been controlled without the 'lightning' calculations of computers, monitoring the descent by taking in information about the position, speed and direction, and giving out information – and acting upon it – about the rocket 'burns' which were necessary. How strange that the physical principles and the mathematical equations involved were those formulated by Newton, almost without embellishment!

Laws, Hypotheses and Theories

It is very easy in discussing the foundations of physics to try to introduce distinctions between the terms 'law', 'hypothesis' and 'theory' which are not really of great significance. The main truth about physics, experimental and theoretical, is that its direction is always towards the discovery of new facts and the arrangement of new and old facts in as simple a scheme as possible. When it is possible to express in a single statement or formula the key to a large number of experimental situations it is common to attach to it the status of a 'physical law'. We have already had an example of a formula ($v = gt$) which covers a large number of situations in that it tells us how the velocity of fall of *any* body towards the earth would increase with respect to the time of fall. However, this might not be thought of sufficient general significance to qualify as a physical law. A more general statement would be one which defined how bodies accelerated when under the action of *any* force, not necessarily gravitational, and not necessarily steady; this is, of course, Newton's second law of motion which has already been discussed.

The great conciseness of mathematical statements makes them ideally suited to the formulation of physical laws, and very many of the laws of physics are expressed in mathematical form. It would be a mistake, however, to suppose that only mathematical statements are important. It is in an equal sense a law of physics that electrons have been identified to have many properties typical of particles, and this statement can be made without mathematical symbolism.

It is also misleading – although it rarely leads to any worse effects than an expenditure of time upon philosophical or metaphysical discussion – to suppose that the use of the word 'law' implies compulsion. The scientist would normally consider the significance of this term to be simply that which has here been given: a physical law is a concise summary of the information derived from a multitude of experimental observations in which some common feature is picked out.

But, the philosophically inclined reader might insist, what happens if on some occasion the law is not 'obeyed'? Can any law be obeyed without exception unless there is some compulsion? The scientist might try to disregard the second question. To the first he would probably say: if a 'law' is not obeyed, even on a single occasion, it is no law. He would wish to check in detail the circumstances surrounding the exceptional observation in order to make sure there had been no error, and would probably repeat the experiment many times. This would be partly in order to see whether there might be another failure to obey the law, but more important, to give him an opportunity to watch for any intruding element in the experiment which might affect the result. As a matter of fact, it has often been in situations of this sort that new discoveries have been made in physics. The intrusion is often the most interesting thing in the experiment.

It may sometimes happen, when an experimental situation is carefully examined in such a way, that the accepted relevant laws are found to be nearly but not quite correct. This is a particularly interesting state of affairs. A *systematic* departure from the accepted law suggests that some finer detail remains to be discovered. A very important example was the discovery, frequently discussed in this book, that mass and energy are interchangeable. For a long time it had been accepted that energy was conserved – that is, it was indestructible. If it disappeared in one form it must re-appear in another form. For example, when the kinetic energy of the water falling over a waterfall disappears at the bottom, it re-appears as heat energy and the water rises slightly in temperature. This general principle was known as the law of conservation of energy. Equally, it was thought to be obvious that matter was indestructible and therefore that *mass* must always be conserved. It is now known, of course, that under certain circumstances, mass can be converted into energy, and vice versa. So the law has to be altered or re-stated. We must either retain the law of conservation of energy with the stipulation that mass is one of the forms of energy now to be considered, or perhaps state a new law of the conservation of mass-energy. In either case we shall certainly have to 'repeal' one law and lay down another. This does not worry the scientist as much as it might worry the philosopher who has been following his work. In science, the experimental fact takes first place.

There is another lesson to be learnt from the example of the waterfall. For where does the kinetic energy of the water come from in the first place? Of course, we know that the water is accelerated because of the force of gravity. But in order to allow the law of conservation of energy a general validity we have to invent a new form of energy and say that the 'potential' energy of the water at the top of the waterfall has been converted into kinetic energy at the bottom. This would be called gravitational potential energy; other examples of potential energy (elastic energy in these cases) would be the hidden energy in, say, a compressed valve spring, or the stretched rubber of a catapult, which could at any time be converted into kinetic energy.

The term 'hypothesis' is usually used in physics to refer to assertions of the same kind as those made in stating physical laws, but thought at the time to be provisional, or temporary, in character. In other words, a hypothesis is an intelligent guess. The results of further experiments would be expected, ultimately, either to confirm hypotheses and give them permanent status as laws, or to dispose of them as unprofitable or useless.

The term 'theory' is not always used with any precise intention. This word is very commonly used in connection with rather large-scale mathematical treatments, such as exist in the theory of relativity or the quantum theory – to name two theories which are well known by name – and cover a large number of experimental situations. However, it is also used almost as a synonym for 'hypothesis'. It is often somewhat misused by students, who might say, perhaps after an unsuccessful experiment, '*Theoretically*, this should have behaved quite differently.'

A certain impatience with too much discussion about the philosophical status of laws, hypotheses and theories, has become a characteristic of modern science. This is not necessarily, as non-scientists often assert, additional evidence of the shallowness of the scientifically educated, The point of view has in itself a real importance in science as we have seen in discussing the impact of recent theories upon the mode of thought of physicists.

Bibliography

INTRODUCTION

BECK, S. D., *The Simplicity of Science*, Penguin Books, 1962.

PRICE, D. J. DE SOLLA, *Little Science, Big Science*, New York, Columbia University Press, 1965.

RATTRAY TAYLOR, G., *The Doomsday Book*, Thames and Hudson, 1970.

ZIMAN, J., *Public Knowledge: The Social Dimension of Science*, Cambridge University Press, 1968.

CHAPTER 1

HECKMAN, H. H. and STARRING, P. W., *Nuclear Physics and the Fundamental Particles*, Holt, Rinehart and Winston, 1967.

HUGHES, I. S., *Elementary Particles*, Penguin Books, 1972.

MUIRHEAD, H., *The Physics of Elementary Particles*, Pergamon, 1968.

PAUL, E. B., *Nuclear and Particle Physics*, North Holland Publishing Company, 1969.

POWELL, C. F., FOWLER, P. H. and PERKINS, D. H., *The Study of Elementary Particles by the Photographic Method*, Pergamon, 1959.

CHAPTER 2

BARBER, M., *Induced Radioactivity*, North Holland Publishing Company, 1969.

BURCHAM, W. E., *Nuclear Physics*, Longmans, 1967.

CHADWICK, J., *Radioactivity and Radioactive Isotopes*, Pitman, 1961.

DZHELEPOV, B. S. and PEKER, L. K., *Decay Schemes of Radioactive Nuclei*, Pergamon, 1961.

INTERNATIONAL CONFERENCE ON ATOMIC MASSES, Winnipeg, University of Manitoba, 1967.

MATTHEWS, P. T., *The Nuclear Apple*, Chatto and Windus, 1971.

REID, J. M., *The Atomic Nucleus*, Penguin Books, 1972.

CHAPTER 3

BUTTLAR, H. VON, *Nuclear Physics; An Introduction*, Academic Press, 1968.

ENGE, H. A., *Introduction to Nuclear Physics*, Addison-Wesley, 1966.

GIBSON, W. H., *Nuclear Reactions*, Penguin Books, 1971.

LIVINGSTON, M. S. and BLEWETT, J. P., *Particle Accelerators*, McGraw-Hill, 1962.

PERSICO, E., FERRARI, E. and SEGRE, S. E., *Principles of Particle Accelerators*, Benjamin, 1968.

WEINSTEIN, R., *Interaction of Radiation with Matter; Nuclear Engineering Fundamentals*, book III, McGraw-Hill, 1964.

CHAPTER 4

DELCROIX, J. L., *Plasma Physics*, vols. I and II, John Wiley, 1965.

FREMLIN, J. H., *Applications of Nuclear Physics*, English Universities Press, 1964.

GLASSTONE, S., *Source Book on Atomic Energy*, Van Nostrand, 1967.

GREEN, H. S. and LEIPNIK, R. B., *Sources of Plasma Physics*, Wolters-Noordhoff, 1970.

HETRICK, D. L., *Dynamics of Nuclear Reactors*, University of Chicago Press, 1971.

MESSEL, H. and BUTLER, S. T., *Nuclear Energy Today and Tomorrow*, Heinemann Educational Books, 1971.

ZYSIN, U. A., LBOV, A. A. and SEL'CHENKOV, L. I., *Fission Product Yields and their Mass Distribution*, Consultants Bureau, New York, 1964.

CHAPTER 5

GOTTLIEB, M., GARBUNY, M. and WALKER, W. E., *Seven States of Matter*, New York, Walker and Co., 1966.

HOLDEN, A. and SINGER, P., *Crystals and Crystal Growing*, Heinemann, 1961.

KRATOCHVIL, P., *Crystals*, Iliffe Books, 1967.

CHAPTER 6

CROPPER, W. H., *The Quantum Physicists*, Oxford, 1970.

KITTEL, C., *Introduction to Solid State Physics*, New York, John Wiley, 1956.

MACDONALD, D. K. C., *Near Zero*, Heinemann, 1962.

PEIERLS, R., *The Laws of Nature*, George Allen and Unwin, 1955.

BIBLIOGRAPHY

REICHENBACH, H., *The Rise of Scientific Philosophy*, Cambridge University Press, 1954.

ROTHMAN, M. A., *The Laws of Physics*, Penguin Books, 1966.

CHAPTER 7

HARRIS, D. J. and ROBSON, P. N., *Vacuum and Solid State Electronics*, Pergamon, 1963.

HOLLINGDALE, S. H. and TOOTHILL, G. C., *Electronic Computers*, Penguin Books, 1970.

TOLANSKY, S., *Revolution in Optics*, Penguin Books, 1968.

CHAPTER 8

BECK, W. S., *Modern Science and the Nature of Life*, Penguin Books, 1961.

CHAPMAN, D. and LESLIE, R. B., *Molecular Biophysics*, Penguin Books, 1967.

CLOWES, R., *The Structure of Life*, Penguin Books, 1967.

CHAPTER 9

ABETTI, G., *The Sun*, trans. J. B. Sidgwick, Faber and Faber, 1963.

BERLAGE, H. P., *The Origin of the Solar System*, Pergamon, 1968.

GAMOW, G., *A Star Called the Sun*, Penguin Books, 1967.

MOTZ, L. and DUVEEN, A., *Essentials of Astronomy*, part 1, Blackie and Son, 1966.

WHIPPLE, F. L., *Earth, Moon and Planets*, Cambridge, Mass., Harvard University Press, 1963.

CHAPTER 10

BOK, B. J. and BOK, P. J., *The Milky Way*, Cambridge, Mass., Harvard University Press, 1957.

MOTZ, L. and DUVEEN, A., *Essentials of Astronomy*, parts 2 & 3, Blackie and Son, 1966.

UNSÖLD, A., *The New Cosmos*, trans. W. H. McCrea, part 2, Longmans, 1969.

CHAPTER 11

MCVITTIE, G. C., *Fact and Theory in Cosmology*, Eyre and Spottiswoode, 1961.

MOTZ, L. and DUVEEN, A., *Essentials of Astronomy*, part 4, Blackie and Son, 1966.

UNSÖLD, A., *The New Cosmos*, trans. W. H. McCrea, part 3, Longmans, 1969.

306

CHAPTER 12

BONNOR, W. B., *The Mystery of the Expanding Universe*, Eyre and Spottiswoode, 1965.

HEY, J. S., *The Radio Universe*, Pergamon, 1971.

KAHN, F. D. and PALMER, H.P., *Quasars*, Manchester University Press, 1967.

MCVITTIE, G. C., *Fact and Theory in Cosmology*, Eyre and Spottiswoode, 1961.

MEADOWS, A. J., *Stellar Evolution*, Pergamon, 1967.

MOTZ, L. and DUVEEN, A., *Essentials of Astronomy*, part 4, Blackie and Son, 1966.

SCIAMA, D., *Modern Cosmology*, Cambridge University Press, 1971.

WEEKES, T. C., *High-Energy Astrophysics*, Chapman and Hall, 1969.

APPENDIX

FEYNMAN, R., *The Character of Physical Law*, B.B.C., 1965.

Index